OAT
BRAN

Edited by

Peter J. Wood

Centre for Food and Animal Research
Agriculture Canada
Ottawa, Ontario, Canada

Published by the
American Association of Cereal Chemists, Inc.
St. Paul, Minnesota, USA

Cover photograph: Graphics Section, Agriculture Canada

Library of Congress Catalog Card Number: 92-75520
International Standard Book Number: 0-913250-77-5

©1993 by the American Association of Cereal Chemists, Inc.

Printed in the United States of America on acid-free paper

American Association of Cereal Chemists
3340 Pilot Knob Road
St. Paul, Minnesota 55121-2097, USA

Contributors

James W. Anderson, Metabolic Research Group, VA Medical Center, University of Kentucky College of Medicine, Lexington, Kentucky, USA

Susan R. Bridges, Metabolic Research Group, VA Medical Center, University of Kentucky College of Medicine, Lexington, Kentucky, USA

R. Gary Fulcher, Department of Food Science and Nutrition, University of Minnesota, St. Paul, Minnesota, USA

Marvin K. Lenz, John Stuart Research Laboratories, The Quaker Oats Company, Barrington, Illinois, USA

Judith A. Marlett, Department of Nutritional Sciences, University of Wisconsin-Madison, Madison, Wisconsin, USA

S. Shea Miller, Centre for Food and Animal Research, Research Branch, Agriculture Canada, Ottawa, Ontario, Canada

David Paton, Agriculture Canada, Crop Utilization Research Unit, POS Pilot Plant Corporation, Saskatoon, Saskatchewan, Canada

Fred L. Shinnick, John Stuart Research Laboratories, The Quaker Oats Company, Barrington, Illinois, USA

Peter J. Wood, Centre for Food and Animal Research, Research Branch, Agriculture Canada, Ottawa, Ontario, Canada

Foreword

In 1989 the public appetite for oat bran was at its peak. Both the product itself and media reports describing miraculous health benefits were avidly consumed. Writing about oat bran seemed as big a business as manufacturing it. As a result, there was an unprecedented demand, which, combined with unfavorable weather, led to a shortage of suitable milling-grade oats.

Many products were promoted with the phrase "contains oat bran" and carried a tacit implication of health benefit, particularly reduction of serum cholesterol and of coronary heart disease risk. Unfortunately nobody had any clear idea what this product oat bran was or should be, nor for that matter who should be eating it or how much.

Although the major manufacturers aimed for a quality product, not all were as scrupulous, and there was certainly some misleading marketing. Even where this was not deliberate, there was a patent absurdity about some of the implications, for example that consuming muffins that were high in fat and sugar but contained little oat bran could be of specific benefit. Nutritionists and regulators were concerned that the two words "oat bran" were considered sufficient to sell the product rather than the product's nutrient content, and that the primary message—to reduce fat and calorie intake and increase dietary fiber—was being lost in the oat "hype." Cartoonists and comedians are always quick to spot absurdity, providing yet another spin-off from the oat bran industry. A "Cathy" cartoon, for example, commented that the average North American was too busy to realize that there was not enough time in a day to eat all the required muffins and then burn off all the calories consumed.

In my laboratories, we analyzed "Oat Bran Tablets," which contained 3.5% ß-glucan. This is the component of interest to the prospective purchaser of these tablets. We found that a daily dose of about 35 is required to reach the level of a serving of rolled oats and 250–300 to match the dose used by J. W. Anderson in his early oat bran studies (Chapter 6).

In an effort to introduce some order into this situation, the American Association of Cereal Chemists convened a committee that adopted a tentative definition for oat bran (Chapter 1), which has achieved a reasonable level of consensus. In addition, it was suggested that a book be compiled that would attempt to describe the nature of oat bran, its means of manufacture and properties, and what was known about its physiological effects. The present volume is the result. As editor, and on behalf of the contributors, I hope that as much detailed technical and scientific information on oat bran as is available at the time of writing can be found in these pages, or that it can be found in the literature cited. The focus of this monograph is oat bran, but frequently data dealing with the whole oat seed are relevant. To some extent, this volume may be considered an update of our parent text (*Oats: Chemistry and Technology*[1]), where much background data dealing with oats in general may be found.

The majority of the contributors to the book selected themselves by virtue of a long-standing interest in oats—an interest that, like my own, may not initially have been motivated by the potential health benefits of increased intake of dietary fiber. The attention we have paid professionally to oats, or dietary fiber, implies at least a minimum level of belief that the product has some commercial and nutritional value. It is thus difficult to avoid appearing promotional, although this is not the intent of this book. Nevertheless, in my opinion, the evidence supports the view that oat bran is a useful source of complex carbohydrate in the diet and has a specific effect attributable to the soluble fiber, or ß-glucan. The magnitude of this effect and the population that might benefit have not been established. These questions are not easily resolved. The necessary research may be time-consuming and expensive, and in some respects lacks the appeal of being on the "cutting edge." A generally accepted protocol for such studies would be of benefit. The ultimate goal must be a better understanding of mechanism.

The publication of an article by Swain et al in January, 1990,[2] resulted in a precipitous fall from grace for oat bran. The reasons behind the instant, and apparently delighted, acceptance by the media of this article as the definitive oat bran research would make an interesting study in itself. Indeed, the industry and public might benefit from a detailed examination of the whole "oat bran story," which seems to exemplify the essential difficulties involved with nutrition and health claims and also encompasses the question of media competence in reporting these and other complex scientific issues. Offensively, but all too typically, most of this media event appeared to proceed without regard for the sensibilities of the population at risk—those with either potential or established coronary heart disease related to elevated serum cholesterol levels.

The article of Swain et al (1990) was not a surprising or pioneering study but dealt expertly with a controversy that already existed and that has not yet been resolved. The article has been variously criticized (Chapter 6), and recent data[3] suggest that the subject group used (young females with normal serum cholesterol levels) was unlikely to respond to oat bran. It is my hope that our own work with the putative active component of oats, the ß-glucan, and other studies of the effects and mechanism of action of oat bran and dietary fiber, might contribute to resolving this controversy. Regardless of the outcome of such studies, events of the last decade have stimulated new ways of looking at oats as a food ingredient. It is no longer dismissed on the basis of poor baking characteristics, and new processes are being developed and useful functionalities identified (Chapter 2).

Oat bran is unlikely to receive disproportionate attention again. It is perhaps as well that this book, aimed at the scientific community, should finally appear in today's less volatile environment.

P. J. Wood
June 25, 1992.

[1]F. H. Webster, ed. 1986. Oats: Chemistry and Technology. Am. Assoc. Cereal Chem., St. Paul, MN.
[2]Swain, J. F., Rouse, I. L., Curley, C. B., and Sacks, F. M. 1990. Comparison of the effects of oat bran and low-fiber wheat on serum lipoprotein levels and blood pressure. New Engl. J. Med. 322:147-152.
[3]Ripsin, C. M., and Keenan, J. M. 1992. The effects of dietary oat products on blood cholesterol. Trends Food Sci. Technol. 3:137-141.

Contents

OAT BRAN

Chapter 1

Structure of Oat Bran and Distribution of Dietary Fiber Components

R. Gary Fulcher
Department of Food Science and Nutrition
University of Minnesota
St Paul, Minnesota 55108, USA

S. Shea Miller
Centre for Food and Animal Research
Agriculture Canada
Ottawa, Ontario K1A 0C6, Canada

Introduction

The term "bran" is applied to a range of products derived from cereal grains and usually relates specifically to the outer layers of the grain or caryopsis. Wheat bran, for example, is a product of commercial roll-milling that is obtained by successive scraping of attached endosperm materials from the outer layers of the grain. The final product is almost devoid of any attached endosperm tissue and is characterized by having high levels of insoluble fiber, ash, vitamins, lipid, and pigments and, in high-moisture circumstances, high levels of hydrolytic enzyme activity. This is a simplistic but common view of a typical "bran," and we have come to think of brans from other grains as being similar. Certainly, the outer layers of cereal grains share many similar morphological features, including aleurone layer, pericarp, nucellus, and seed coat (testa), and the differences between these outer layers lie primarily in relatively small morphological and chemical differences. (For instance, barley aleurone layers invariably are two to three cells thick, whereas those of oats and wheat are

1

single-celled; wheat contains high levels of aleurone niacin, whereas the level in oats is relatively low). Most cereals have bran tissues enriched in lipid, protein, and minerals.

Although many relatively hard-textured cereals (e.g., wheat, barley, corn, sorghum) can be dry-milled to some extent in roller or abrasive mills to produce concentrates of outer grain tissues (i.e., "bran"), the soft texture and usual high lipid levels of the oat kernel preclude such manipulations, as any miller will attest. The endosperm "flour" does not easily separate from the outer layers, and the lipids contribute to a product that is not easily sieved. (Alternative approaches to producing "clean" endosperm and bran with wet-milling in water or other solvents have been developed. However, these processes are not in common use, although they offer considerable potential for production of oat fractions with commercially important properties [see Chapter 2]).

Because the endosperm tissues in the oat grain do not separate cleanly and efficiently from the outer fibrous layers, typical commercial oat bran products contain large amounts of adhering starchy endosperm. Consequently, any discussion of oat "bran" (at least in commercial or nutritional terms) must include consideration of these endosperm properties. In addition, increased interest in the nutritional merits of oat bran dictates that emphasis must be placed on those attached endosperm components that appear to be primary contributors to improved serum lipid profiles, namely the mixed-linkage $(1\rightarrow3),(1\rightarrow4)$-ß-D-glucan that occurs in abundance in endosperm cell walls (see Chapter 4). Indeed, the following definition of oat bran (AACC, 1989) clearly identifies the importance of the adhering endosperm tissues in typical oat bran products:

> Oat bran is the food which is produced by grinding clean oat groats or rolled oats and separating the resulting oat flour by sieving, bolting and/or other suitable means into fractions such that the oat bran fraction is not more than 50% of the starting material, and has a total β glucan content of at least 5.5% (dry weight basis) and a total dietary fiber content of at least 16.0% (dry weight basis), and such that at least one-third of the total dietary fiber is soluble fiber.

Because the starchy endosperm is the primary site of ß-glucan deposition, it is apparent that this definition is based on the expectation that commercial oat bran will contain significant levels of the starchy tissues. Since the aleurone and other outer layers account for considerably less than 50% of the groat, the mill yield included in the definition also reflects the high levels of starchy endosperm expected in the bran.

In an earlier discussion of oat kernel structure (Fulcher, 1986), the organization of the groat represented in Figure 1 was described in detail, with considerable emphasis on the microchemical organization (i.e., cellular structure and function relationships) of the outer layers of the grain (aleurone layer, pericarp, testa, nucellus, etc.). While there have been important recent additions to our understanding of the chemistry of some oat components (e.g., endosperm cell wall polymers [Wood et al, 1991a], starch [Paton, 1986], grain phenolics [Collins, 1986, 1989; Collins et al, 1991]) and of the changes that they undergo during cooking (Yiu et al, 1991), there has been little additional insight into the morphological organization of the bran and associated components. Recently, however, we have undertaken a number of morphological and biochemical analyses relating to variation in the distribution of polysaccharides in the endosperm cell walls. These polysaccharides occur in relatively high concentrations in commercial oat bran and significantly influence nutritional quality (see Chapters 5 and 6). Results of our analyses indicate that ß-glucan in particular is quite variable from variety to variety, not only in overall concentration and perhaps molecular configuration, but also in the manner of association with the bran. In the following paragraphs, some of the distributional characteristics of mixed-linkage ß-glucan are described and compared among different oat varieties and other genera of cereals. Quantitative imaging methods used in these analyses are described briefly.

Methods for Examination of Groat Structure

GRAIN MICROSCOPY AND MICROCHEMISTRY

Although many different forms of microscopy have been used to examine cereal structure and composition for over a century (O'Brien, 1983), we have relied on fluorescence microscopy as a primary method for component visualization for over 20 years. The reasons for selecting this approach are several and have been discussed elsewhere (Fulcher et al, 1989). Briefly, however, it bears repeating that the vast majority of cereal structures are well within the resolving power of the light and fluorescence microscopes and, at the very least, the fluorescence microscope provides resolution, chemical specificity, and sensitivity rarely matched by other forms of optical microscopy. In addition, instruments designed for quantitative imaging (e.g., digital image analysis and scanning microspectrophotometry) are ideally suited to precise characterization of fluorescence images.

The fluorescence microscope is an extremely powerful instrument for chemical analysis, and it is this use that allows us to evaluate the composition, concentration, and distribution of a wide range of polymers and low-molecular-weight materials in the oat kernel. Methods for microscopic detection of many important constituents of oats and other cereals have been described (Fulcher and Wong, 1980; Fulcher et al, 1989). In addition to detection of naturally fluorescent materials such as phenolic acids, specific stains and fluorescent probes may be applied to a range of sectioned oat kernels and/or products for visualization of particular components. These are summarized in Table 1 (compare Fulcher, 1986).

Detection of these several components requires the use of a relatively simple fluorescence microscope equipped with fluorescence filters of the type already described (Fulcher et al, 1989). Because fluorescent substances are characterized by having specific excitation and emission spectra, successful fluorescence microscopy is critically dependent upon the use of filter systems having filters that 1) excite fluorescent substances in a specimen at appropriate wavelengths (exciter filter) and 2) subsequently filter out all but the emitted fluorescent wavelengths from the final image (barrier filter).

Although we have found the filters described earlier to be suitable for a wide range of applications in cereal analysis, many other filter systems may also be used, as may many other stains and probes. The primary requirement is simply that the instrument be capable of providing high-intensity excitation of naturally fluorescent material or applied stains at or near the appropriate peak wave-lengths.

By restricting discussion of groat microstructure to fluorescence techniques, we in no way imply that these are the only approaches that are suitable. Indeed, it is important to emphasize that complete and appropriate analysis of cereal structure and composition requires the application of a range of microscopic methods (e.g., transmission and scanning electron, bright-field, and interference contrast microscopy) in combination with appropriate chemical and physical analyses. Our experience has been simply that the fluorescence approach allows more precise in situ definition of chemical constituents in routine analyses. This is due, in part, to improved instrumentation over the past 20 or so years and particularly to the dramatic increase in the

Figure 1. Diagram showing the anatomical components of a typical oat kernel (*Avena sativa* L.). The area of the kernel including the germ (from the inset marked C to the lower tip of the kernel) is referred to in the text as the proximal region of the kernel. The inset marked B is in the central region of the kernel, and the inset marked A is in the distal region of the kernel. The cross-section at the lower right of the figure shows an area through the proximal region, containing some germ tissue and equivalent to material used in microspectrophotometric "mapping" of ß-glucans. A–C are higher magnifications of portions of the bran, central starchy endosperm, and germ, respectively. (Reprinted, with permission, from Fulcher, 1986)

numbers and specificities of fluorescent probes during the same period (e.g., Haugland, 1989). For detailed discussions of fluorescence applications in cereal research, see Munck (1989), Guilbault (1989), Barnes and Fulcher (1989), and Fulcher et al (1989).

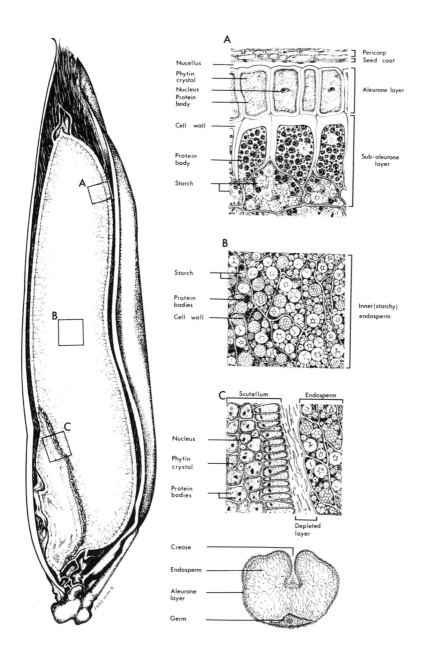

SCANNING MICROSPECTROPHOTOMETRY

In 1975, one of us (RGF) showed (as reported in Hughes and McCully [1975]) that Calcofluor White M2R New had a remarkable affinity for the endosperm cell walls in barley (see Fincher [1975] for a discussion of barley endosperm cell wall composition). In several subsequent studies it has become quite clear that this and similar dyes could be used in two related and important ways. First, they were capable of selective staining and visualization of cereal ß-glucans in situ (Wood et al, 1983). Second, their highly sensitive interaction with ß-glucans in solution (Wood, 1980) allowed routine use in rapid detection systems such as flow injection analysis (e.g., Aastrup, 1988; Jørgensen 1988; Jørgensen and Aastrup, 1988a,b). Both of these methods (fluorescence microscopy and flow injection analysis) are in common use and, in addition to the enzymatic ß-glucan assay defined by McCleary and Glennie-Homes (1985), they form a significant part of the analytical arsenal that is being applied routinely to analysis of oat ß-glucan (see also Chapter 4).

TABLE 1
Methods for Fluorescence Microscopic
Detection of Cereal Constituents

Component	Method	Reference
Protein	Acid fuchsin 8-Anilo-naphthalene sulfonic acid	Fulcher and Wong (1980) Gates and Oparka (1982)
Mixed-linkage ß-glucans	Calcofluor Congo red	Wood and Fulcher (1978) Fulcher and Wood (1983)
Phenolic acids	Autofluorescence	Fulcher et al (1972a) Fincher (1976)
Niacin	Cyanide bromide/ p-aminobenzoic acid	Fulcher et al (1981)
Aromatic amines	p-Dimethylamino- cinnamaldehyde	Fulcher et al (1981)
Phytin	Polarizing optics, Acriflavine HCl	Fulcher (1972) Yiu et al (1982), Irving et al (1991)
Starch and periodate- sensitive cell walls	Periodate/Schiffs Periodate/acriflavine HCl	O'Brien and McCully (1981) Fulcher and Wood (1983)
Cell wall polymers	Lectins	Miller et al (1984) Yiu et al (1991)
Lipids	Nile blue A	Hargin et al (1980) Irving et al (1991)

Although these methods dictate that mixed-linkage ß-glucan is now among the most readily analyzed of polysaccharides, there remain several questions regarding the organization of this nutritionally important component in the groat, and particularly in commercial brans. It is not known, for example, whether ß-glucan is structurally similar throughout the endosperm or whether it varies somewhat from one region of the endosperm to another (e.g., as a function of the stage of development at which it was synthesized). Similarly, while processing experience suggests that the total ß-glucan content of a sample of groats is not necessarily indicative of final concentrations in manufactured bran (Wood et al, 1991b), no systematic and quantitative analysis of ß-glucan distribution has been undertaken that would account for these anomalies (although earlier microscopic analyses suggested that considerable variation in distribution patterns was possible [Fulcher, 1986]). Finally, it is also of some significance to assess the structural organization of the endosperm cell walls on a varietal basis to answer questions such as the following: Are endosperm cell walls similar in thickness in domestic and primitive cultivars, and do different thicknesses imply (as expected) differences in digestibility and processing properties? Is the fine structure of the "typical" ß-glucan-rich cell wall similar for all varieties, or do they differ in structural organization? Certainly milling studies on a variety of domestic cultivars suggest that there are also significant *physical* differences among varieties in groat response to dehulling, and these may in part be due to endosperm or bran structure, as well as to differences in gross kernel morphology (Symons and Fulcher, 1988a,b,d). These are but a few of a wide range of questions relating to the structure and organization of oat bran and particularly ß-glucan that remain unanswered. If we are to understand the mechanisms either of synthesis and breakdown in the groat during development and germination, or of digestion in the mammalian gut, further details of structure and organization are necessary. Quantitative imaging using *scanning microspectrophotometry* is a particularly useful method for "mapping" concentrations and distributions of ß-glucan in the groat. Examples illustrated in this chapter show very clearly that varieties differ significantly in the way in which the polymers are distributed.

Essential elements of a scanning microspectrophotometer suitable for analysis of cereal products have been described in detail elsewhere (Fulcher et al, 1989; Irving et al, 1989). Essentially, the instrument (model UMSP80, Carl Zeiss Ltd, Thornwood, NY) used to obtain the results shown in this chapter is a research fluorescence microscope equipped with monochromaters or excitation/barrier filter systems suitable for maximizing fluorescence in Calcofluor-stained specimens to reveal ß-glucan, and a scanning stage that allows scanning across the surface of a specimen to map concentrations of the stained fluorescent polymers throughout each specimen. Measurements are obtained via a photomultiplier situated at an appropriate image plane. The monochromaters, photomultiplier, and scanning stage are all controlled by a 386-MHz IBM-compatible Compaq computer and appropriate software. In this analysis, the specimen-scanning software, MAPS (Carl Zeiss Ltd.), has been used to define the distribution of ß-glucans in the endosperm, with particular emphasis on the ß-glucan likely to be associated with the bran. Although many other materials can be assessed in a similar way (e.g., phenolic acids, protein, or any of the materials detected in Table 1 and in Irving et al [1991]), such studies have not yet been conducted on oats. We would expect, however, that significant differences in the concentration and distribution of physiologically important materials will be detected as more such analyses are conducted.

ß-GLUCAN MAPPING

To "map" ß-glucan distribution in the groat, oat kernels were first dehulled by hand. Samples of known ß-glucan content (cultivars Donald, 3.7%; OA516-2, 4.0%; Tibor, 4.6%; Woodstock, 5.1%; and Marion, 6.4%) were obtained from the Eastern Cooperative Oat Test (Agriculture Canada, Ottawa) and embedded in blocks of polyester resin using the method of Symons and Fulcher (1988c).

To position seeds for embedding, plasticene was rolled flat, and a shallow cardboard box (sides approximately 4 cm long, 1.5 cm deep, open at both ends) was pushed into the surface such that a tight seal was achieved between the cardboard and the plasticene. Kernels were aligned vertically and (at embryo end) pushed lightly into the plasticene. Polyester resin mixed with color (white) and hardener was then poured into the box to completely cover the oats. After the resin hardened, the cardboard was removed, and the plasticene was carefully scraped off.

For microspectrofluorometry of ß-glucan distribution, the blocks were then abraded to the desired depth in the groats using a sander, and the exposed surfaces containing seed cross-sections were polished with fine sandpaper by hand. The surfaces of the blocks were treated for 2 min with 0.01% Calcofluor in 50% ethanol buffered with phosphate (50 mM, pH 8), then rinsed with 50% ethanol. The blocks were then counterstained for 30 sec with Fast Green FCF (0.1% in 50 mM acetate buffer, pH 4.0), rinsed with distilled water, and blotted dry. The exposed surfaces were then scanned at 365 nm using a Zeiss UMSP80 scanning microspectrophotometer.

Before scanning a particular specimen, the UMSP80 photomultiplier was calibrated to optimize fluorescence response and adjusted to a stable level. The specimens were excited at 365 nm, which corresponds to one of several major spectral output bands in the mercury illuminator and which is the wavelength maximum of the excitation spectrum of ß-glucan-bound Calcofluor. Subsequently, specimens were scanned in a matrix fashion across the exposed surface at pre-determined intervals to generate a plot of fluorescence intensities across the sample. Using this approach, several cultivars were examined to define ß-glucan distribution in the groat.

ß-GLUCAN DETERMINATION

To ensure that ß-glucan values obtained by microspectrophotometry were compatible with data obtained by other analytical methods, it was necessary to apply conventional assays for ß-glucan. This is a requirement when any microchemical interpretations are made on the basis of assumed microscopic specificities. Although it has been clear for several years that Calcofluor can be used to localize ß-glucans in oats precisely, there remains the necessity to run parallel chemical assays whenever microscopic analyses are conducted. We therefore determined the gross ß-glucan content of a subsample of the groats used for microspectrophotometry using Miller's (1992) modification of the lichenase digestion method of McCleary and Glennie-Holmes (1985).

Kernel Structure and Location of the Bran

A typical oat kernel is illustrated diagrammatically in Figure 1 and shows the major compartments and cellular elements of the grain. These have been described in detail previously (Fulcher, 1986), but this chapter emphasizes some of the primary properties of the bran, a combination of several tissues.

The commercial oat fraction that is most commonly referred to as *bran* is approximated diagrammatically in Figure 1A, which is an enlargement of the outer layers of the groat (pericarp, seed coat, nucellus, and aleurone layer) plus one to several cell thicknesses of adhering starchy endosperm containing starch, protein, and elevated levels of ß-glucan.

Although the aleurone layer is perhaps the most complex of the bran tissues, depending on processing conditions the adhering starchy endosperm material may be in approximately equal proportions relative to the aleurone and outer layers, or it may provide up to 80% or more of the final product. Obviously, then, a discussion of oat bran must include this subaleurone and endospermic material, most particularly in view of its possible role in influencing blood serum lipid profiles (see Chapters 5 and 6). Unequivocal identification of these juxtaposed elements would improve definition, or understanding, of the term "oat bran."

Despite problems in assessing the amount of adhering endosperm, the bran product we are familiar with can be defined morphologically in relatively simple terms: it is that portion of the groat (dehulled oat) composed primarily of the outer tissue layers (aleurone layer, nucellus, testa, and pericarp) of the caryopsis but containing additional, variable amounts of adhering material from starchy endosperm and germ. In dry-milled products, the adhering material is a major feature of the bran; in wet-milled material, it is considerably less abundant (see Chapter 2). In either case, the primary chemical traits are strongly influenced by the outer layers, and the product typically is enriched in mineral, lignin, phenolic acids, aromatic amines, basic amino acids, and lipids (Fulcher, 1986). Both soluble and insoluble fibers are enriched in typical bran products (see Chapter 3).

Microchemical Characteristics of Oat Bran

The microchemical characteristics of the several bran tissues (especially the aleurone layer) have been described previously in detail (Fulcher, 1986), and additional microscopic and/or structural information since that report has been minimal. Yiu et al (1991) have described some of the cursory changes that occur in oat tissues during processing, and Collins et al (1991) have defined a new group of fluorescent phenolic acids (avenalumic acids) and avenanthramides (Collins, 1989) that one might expect to occur, and possibly be enriched, in bran tissues.

The three outermost tissues of the groat (pericarp, testa, and nucellus, Fig. 1A) are remnants of the ovary. At maturity they are composed primarily of highly insoluble and at least partially lignified polymeric material and appear to be relatively resistant to degradation. These layers have little or no metabolic activity (although microorganisms may infiltrate any crevices and cracks that form in the groat surface during harvesting or handling, resulting in degradative metabolic changes during storage or processing). These tissues contribute significantly to the overall insoluble fiber content of oat bran. Little if any detailed information is available relating to genetic or environmental influences on either structural or chemical properties of these layers in different cultivars. It is also worth noting that the fluorescence properties of these different layers are distinctly different, at least as visualized microscopically. In view of the high degree of adhesion among these layers, separation and selective chemical characterization of each of the layers will be difficult.

The aleurone layer, in contrast, is a very active metabolic tissue, with the primary functions of providing protection to the grain and generating the hydrolytic enzymes (e.g., lipase [Miller et al, 1989]) that are necessary for degradation and mobilization of the endosperm reserves during germination. The layer is typically one cell thick in oats (50–150 μm). A major characteristic of the aleurone layer is a high concentration of fluorescent compounds in the aleurone cell wall, as shown in Figure 2. Although ferulic acid is a major component of the bran tissues, as shown by high-performance liquid chromatography (Miller, 1992), the avenalumic acids described by Collins et al (1991) might also contribute to the fluorescence properties of the cell walls. A relatively thin layer of ß-glucan is also associated with the inner region of the wall (Wood and Fulcher, 1978; Fulcher, 1986).

The cytoplasm of the aleurone layer continues to resist serious chemical analysis, although some of the more salient features can be described. The layer is rich in lipid and contributes some 35–40% of the total lipid in the groat; approximately 40% is triglyceride (Youngs, 1986). Like most cereals, oat bran has a relatively high protein content (Peterson and Brinegar, 1986), which is enriched (at least in the aleurone layer) in basic amino acids. This is also true for other cereals (Fulcher et al, 1972b; Fulcher, 1986). Aleurone proteins represent a unique and distinctive compartmentation in cereal grains. Oat aleurone proteins have received considerably less attention than those in other cereals. These proteins are stored in aleurone grains or protein bodies, distinct spherical structures packed within each aleurone cell (and visible in Fig. 2). As in wheat, each oat aleurone grain is surrounded by small droplets (spherosomes) of lipid-rich material (Hargin et al, 1980).

Within each aleurone grain are two additional, highly compartmentalized, and easily detectable structures: a crystallized phytin deposit (salts of myo-inositol hexaphosphate) and a curious assembly of vitamins and possible aromatic amines sequestered in an additional carbohydrate-containing structure. With appropriate histochemical treatments, the latter shows significant concentrations of niacin (Fig. 3, after reaction with cyanogen bromide and *p*-aminobenzoic acid) and presumed aminophenol (Fig. 4, after reaction with *p*-dimethyl-amino-cinnamaldehyde). These observations are important in light of the processing industry's continuing interest in both the antioxidant activity of oat products in general and the nutritional quality of the bran. The point is perhaps a simple one: the apparent large quantities of ill-defined structures containing physiologically important substances such as niacin and aminophenol beg further research. We might well

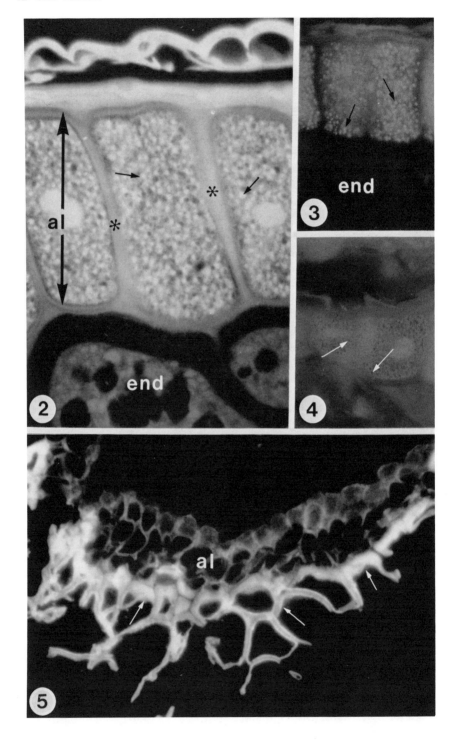

assume that other, as yet uncharacterized, compounds, some of which no doubt also possess metabolic activity, also occur in high concentrations in oat bran.

Distribution of ß-Glucan in the Groat and Bran

As indicated earlier, a key characteristic of oat bran is that, unlike most other cereal brans, it contains large amounts of adhering endosperm, with accompanying significant concentrations of ß-glucan and starch. A typical oat bran fragment and associated ß-glucan is shown in Figure 5. Because the ß-glucan is a cell wall component, it surrounds the major storage polysaccharide, the small compound starch granules that are found in all endosperm cells (Fig. 6). That the adhering starchy endosperm also contains significant protein levels is apparent in Figure 7, which shows the intimate juxtaposition of ß-glucan, starch, and protein. Although both aleurone and endosperm cells were treated with the same reagents in this example, it is apparent that the protein matrix in the two cell types reacts quite differently to fluorescent probes.

These selected micrographs (Figs. 2–7) emphasize that oat bran is not a homogeneous matrix of ß-glucan, protein, and a few nutrients but is rather a highly complex and compartmentalized product that will release its constituents only with difficulty during digestion or processing. The phenolic-acid-rich cell walls of the aleurone layer and the ß-glucan-rich cell walls of the adhering tissues are distinctly different and are major sources of insoluble and soluble fiber, respectively. Each of these unique fiber elements "surrounds" the remaining cytoplasmic components in these tissues, and it can be assumed that they exert a strong influence on the rate and degree of solubilization, extraction, and other modifications that may occur in the gut during digestion and in any commercial process involving physical fractionation and cooking. Recently, Yiu et al (1991) have adapted similar

←

Figures 2–5. Fluorescence micrographs. **2,** View of the outer layers of the groat, including the aleurone layer (al) with autofluorescent phenolic-acid-rich cell walls (*) and numerous aleurone grains (arrows) within each cell. A small amount of associated endosperm tissue (end) is also visible. **3,** Aleurone tissue treated with cyanogen bromide/p-aminobenzoic acid to show high concentrations of niacin reserves (arrows). Note absence of similar reaction products in attached endosperm (end). **4,** Tissue similar to that in Figure 3 but treated with dimethylaminocinnamaldehyde to show supposed aromatic amines in discrete cellular deposits (arrows). **5,** A low-magnification view of a particle of oat bran after treatment with Calcofluor to show ß-glucans (arrows) in the endosperm attached to the aleurone layer (al). (Figures 2–4 adapted from color plates 2, 8, and 9, respectively, in Fulcher, 1986)

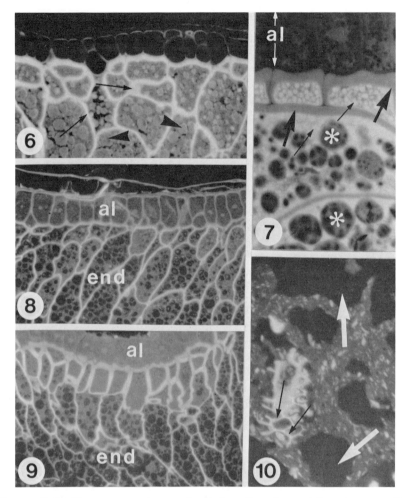

Figures 6–10. Fluorescence micrographs. **6,** A portion of an oat kernel cross-section stained with both periodate/acriflavine to show starch (large arrowheads) and Calcofluor to show the ß-glucan-rich endosperm cell walls (small arrows) surrounding each endosperm cell. **7,** Tissue treated sequentially with periodate/acriflavine (compound starch granules, *), Calcofluor (ß-glucan, large arrows), and acid fuchsin (protein bodies, small arrows). The aleurone layer (al) and underlying endosperm differ in their reactions to histochemical reagents. **8 and 9,** Fluorescence micrographs after Calcofluor staining of cross-sections of the groats of two distinctly different cultivars. Note the relatively uniform staining of the cell walls throughout the endosperm (end) in Figure 8, compared to the quite thick subaleurone cell walls in the cultivar shown in Figure 9. **10,** A Calcofluor-stained section through a commercial puffed oat product, showing the remaining integrity of the ß-glucan-rich cell walls (small arrows) in this type of product. Air cells are clearly visible (large arrows). (Figure 8 adapted from color plate 13 in Fulcher, 1986)

microscopic techniques to evaluate changes imparted by cooking various oat products, but little quantitative information is available relating to the genetic variation influencing the manner in which these materials are sequestered in the groat. Quite clearly, any differences in cell shape or particularly in cell wall thickness (such as that observed in some cultivars of barley [Aastrup and Munck, 1985]) will influence the rate at which these modifications may occur.

In earlier studies, simple microscopic observation suggested that there were distinct patterns of ß-glucan deposition in the groat, depending upon variety. These observations were based on microscopic examination of Calcofluor-stained groat cross-sections, and patterns ranged from unremarkable, relatively uniform distribution throughout the endosperm (Fig. 8), to pronounced concentrations in the subaleurone layer (Fig. 9), the endosperm tissue that lies immediately adjacent to the aleurone layer (Fulcher 1986). (Some of the structural integrity of these ß-glucan-rich cell walls is carried over into highly processed, puffed, oat products [Fig. 10]). To define this variation further, several varieties of oats were assayed for ß-glucan content, and selected kernels were subsequently examined by microspectrophotometric scanning of Calcofluor-stained sections, to "map" the distribution, as described earlier.

ß-Glucan contents of groats are highly variable, ranging in our experience from less than 2% in some primitive varieties to almost 7% in exceptional domestic lines. Samples selected for microspectrophotometry ranged from 3.7 to 6.4% (dry weight basis). Although we have observed that these varieties typically rank in the same order when grown simultaneously in one location, they may range quite widely in final ß-glucan content from year to year or among locations. Details of these comparative studies will be published elsewhere.

Microscopic examination has shown quite clearly that ß-glucan distribution is variable, but simple microscopy seldom allows either a quantitative assessment of these patterns or an overview of an entire cross-section of a groat. Photometric scanning of sections provides solutions to both of these limitations, however, and examples of several such scans are shown in Figures 11–13. Figures 11 and 12 represent comparisons made at three different locations in the groat: 1) at the approximate midpoint of the germ, 2) at the approximate midpoint of the kernel, and 3) approximately one quarter of the distance from the distal end of the groat. A diagrammatic representation of a typical cross-section of the groat that is approximately equivalent to the type of specimen scanned at the first location above is shown in Figure 1 (lower right), and those from the second and third locations would be very similar but lacking germ elements.

The relative fluorescence intensity (RFI) of Calcofluor bound to ß-glucans as measured by scanning microspectrophotometry is represented in the form of intensity profiles. The profiles are oriented such that the kernel crease is positioned at the left of the section (i.e., they are rotated 90° counterclockwise from the position shown in Fig. 1, lower right). In solution, the RFI of bound Calcofluor is approximately proportional to the amount of ß-glucan present (Wood and Fulcher, 1978, 1983; Wood, 1985; Jørgensen, 1988), and for purposes of this discussion, we have assumed that peak heights are proportional to ß-glucan levels at any particular measurement point. This relationship, however, remains to be established unequivocally for material prepared in this fashion.

Figure 11 shows three intensity profiles representing proximal, central, and distal scans of a single kernel of OA516-2, taken from a sample assayed at 4.0% ß-glucan. In the proximal region of the kernel, which contains the embryo (Fig. 11a), the highest concentration of ß-glucan is in the starchy endosperm immediately adjacent to the embryo (subembryo area). This area contains the depleted, or intermediate, layer (see Fig. 1C) and consists of cell walls that remain after the expanding and growing embryo has forced degradative changes in the endosperm tissues in this area of the grain (Fulcher 1986). The cell walls in the embryo itself are very thin and contain little ß-glucan, as indicated by the relative peak heights in the fluo-

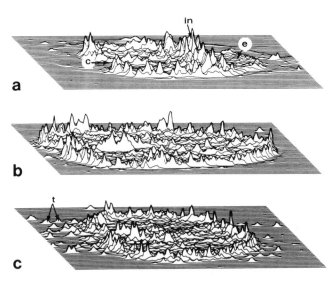

Figure 11. Distribution of ß-glucan in a kernel of OA516-2 oats, detected by scanning the relative florescence intensity (RFI) of bound Calcofluor. **a,** Map of RFI at the proximal end of the kernel (containing the embryo [e]); **b,** map of RFI in the central region of the kernel; **c,** map of RFI at the proximal end of the kernel. in = intermediate layer, c = crease, t = trichomes.

rescence intensity profiles. A relatively high concentration of ß-glucan is also found in the peripheral regions of the kernel, in the subaleurone layer (see Fig. 1). As in the depleted layer, the cell walls in the subaleurone layer are quite thick, and wall thickness decreases in the interior of the starchy endosperm (Fig. 9). Microscopic evidence of thinner walls in the interior of the starchy endosperm was reflected in the lower fluorescence observed in those areas. Differences in cell wall thickness corresponding to differing ß-glucan contents have been reported in barleys with different ß-glucan contents (Aastrup, 1983).

In the central region of the kernel, the very large subembryo deposition is no longer present, and the ß-glucan distribution is similar to that seen in the ventral portion of the proximal section. The ß-glucan content is highest in the subaleurone layer (around the periphery of the kernel and up into the crease), with lower levels in the interior of the section. The distribution in the distal portion of the kernel is the same as in the central region in this variety. Small points of fluorescence are also visible outside the kernel at the distal end. These are caused by autofluorescence of the trichomes, or hairs, attached to the kernel. This autofluorescence measures less than 5% RFI, which is the lower threshold set for data acquisition on the microspectrophotometer, and thus does not interfere with the fluorescence measurement of ß-glucan.

ß-Glucan distribution is rather different in the high-ß-glucan variety Marion (Fig. 12), which had a ß-glucan content of 6.4% for the sample illustrated in this chapter. The greatest difference is observed in the central region of the kernel (Fig. 12b), where the fluorescence intensity is high throughout the section and no clear subaleurone concentration of ß-glucan is apparent. The concentration of ß-glucan is slightly lower toward the dorsal side of the kernel. A relatively even distribution throughout the kernel is also observed in the distal portion of the kernel (Fig. 12c). In the proximal region (Fig. 12a), the distribution of ß-glucan is similar (near the embryo) to that seen in OA516-2 (Fig. 11), although the relative amounts are higher.

Figure 13 compares scans at the mid region of groats of all five cultivars used in the study. These representative views indicate that the ß-glucan is highly concentrated adjacent to the aleurone layer (i.e., the subaleurone layer), at least in the groats containing the lower levels. In most cases, there is also an abundance of the material immediately adjacent to the front face of the scutellum (the intermediate layer, not shown), similar to that visualized in Figures 11a and 12a.

The mean percent RFI of 25 individual scans of the mid-kernel section from the five cultivars of oats correlated highly ($r = 0.97$) with the ß-glucan content of the cultivars determined enzymatically.

Although only a limited number of samples was used, the data suggest that simple measurement of a section of groat from the mid-kernel region is a reasonable indicator of total ß-glucan content. The prospect exists for using a similar approach for single-seed evaluation of the polysaccharide in breeding programs and genetic studies. Full-scale scanning microspectrophotometry would not be necessary; a simple photomultiplier would suffice. To warrant this approach for ß-glucans or any other component, the current data will require expansion to establish the observed high correlation. In practice, the distal portion of kernels of interest could be removed and scanned, and the remainder of the seed, with its embryo intact, would then still be viable and suitable for transplantation to suitable growth media.

Although the data shown in Figure 13 represent only five cultivars of oats, a trend was apparent wherein the high subaleurone concentration of ß-glucan became a less distinctive characteristic as the total ß-glucan content of the cultivars increased. The lack of a distinct subaleurone concentration of ß-glucan in the kernel may reflect differences in cell wall thickness in this area for some cultivars, as noted earlier (Fulcher, 1986), or it may be that a higher concentration

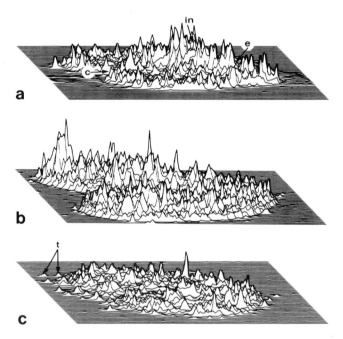

Figure 12. Distribution of ß-glucan in a kernel of Marion oats, detected by scanning the relative florescence intensity (RFI) of bound Calcofluor. **a,** Map of RFI at the proximal end of the kernel (containing the embryo [e]); **b,** map of RFI in the central region of the kernel; **c,** map of RFI at the proximal end of the kernel. in = intermediate layer, c = crease, t = trichomes.

of ß-glucan in the central endosperm (for example, as in Marion) obscures elevated subaleurone concentrations in these cultivars. When thin sections of the central region of both OA516-2 and Marion are examined microscopically, the differences in ß-glucan content in the interior of the starchy endosperm appear to reflect differences in cell size, rather than in wall thickness (not shown). In a limited number of evaluations of OA516-2, the cells in the starchy endosperm were often quite large, whereas in Marion the cells were much smaller, leading to a higher proportion of cell wall per unit of area in Marion than in OA516-2 and possibly explaining the higher whole-grain content of ß-glucan in Marion. It is not known at this time

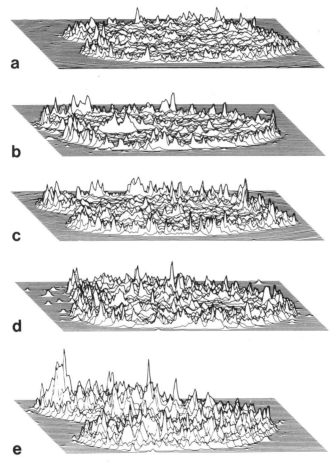

Figure 13. Comparison of ß-glucan distribution in the central region of five cultivars of oats by means of intensity profiles showing distribution of bound Calcofluor. **a,** Cultivar Donald (3.7% ß-glucan); **b,** OA516-2 (4.0% ß-glucan); **c,** Tibor (4.6% ß-glucan); **d,** Woodstock (5.1% ß-glucan); **e,** Marion (6.4% ß-glucan).

whether the apparent differences in cell size in the starchy endosperm relate to ß-glucan differences in other cultivars, nor have the cell sizes in different varieties been compared quantitatively in any strict manner.

Differences in the thickness of cell walls in the subaleurone layers of oats have clear relevance to the production and nature of oat bran. Wood et al (1991b) reported an average enrichment factor of 1.5 in ß-glucan content between whole groats and brans from a selection of oats. Significant changes in ranking, however, were observed in the ß-glucan contents of kernels and brans. The differences in distribution observed in the present study provide an explanation for these observations and suggest that scanning microspectrophotometry may be a valuable tool for selection of cultivars of oats appropriate for production of high ß-glucan brans. In addition, results are rapid and relatively simple to obtain.

For comparison, ß-glucan distribution in barley kernels has also been studied, and these data will be submitted for publication elsewhere. Briefly, however, the distribution of ß-glucan in barley mimics that of high-ß-glucan oat cultivars, regardless of the ß-glucan level. That is, in a limited set of cultivars spanning the range from 3 to 11% ß-glucan, there appears to be little or no particular concentration in the subaleurone region, and the highest levels of polymer are consistently in the center of the kernel (Fig. 14). In wheat, which commonly contains only minimal amounts of ß-glucan (Beresford and Stone, 1983), the highest concentration is invariably in the subaleurone layer, with little or no material throughout most of the starchy endosperm. We are currently continuing these studies to evaluate the diversity of storage patterns in different members of the cereal Gramineae.

Concluding Comments

It is noteworthy, perhaps, that discussions of nutritional quality of oat products rarely include descriptions of ß-glucan content or of the many other elements found in the aleurone layer and associated tissues. This is somewhat surprising in view of the rather extensive

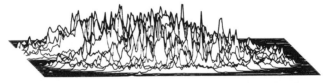

Figure 14. Intensity profile of Calcofluor-stained cross-section of a single barley kernel, showing the high levels of ß-glucan in the central region of the kernel. This specimen was selected from a sample of grain with approximately 11% ß-glucan (dry weight basis).

literature that has sought to relate the biochemistry of oat bran to purported nutritional benefits (see Chapters 5 and 6). In most cases, the ß-glucan content of the material used has not been reported, nor have additional data been provided that might relate to degree of dissociation of bran elements during cooking, particle size, etc.

It is also obvious that the complexity of oat bran, and other cereal brans, requires careful consideration if we are to draw valid conclusions regarding its role in either nutritional or processing contexts. The level of antioxidant(s) commonly found in the tissues is quite remarkable in some cases, but with the exception of the detailed research of Collins and co-workers (Collins, 1986, 1989; Collins et al, 1991), there seems to be little activity in this area. Similarly, dietary fiber assays will remain equivocal until such time as concerted efforts are made to fully identify the polymers associated with the bran. It is perhaps most remarkable that our best definitions of cereal materials continue to relate to starchy endosperm, yet in almost all cereals, the bran and outer envelope contribute a minimum of 20% of the dry weight of the kernel. As evidence for the nutritional benefits of brans increases, confirming the view that whole grains provide distinct advantages over highly refined or extracted products, so too will the requirement for increased biochemical data on both structure and functional relationships in grains.

ACKNOWLEDGMENTS

We are grateful to Vernon D. Burrows, Plant Research Centre, Agriculture Canada, for provision of plant materials and assistance in growing selected cultivars for analysis, and to P. J. Wood, Food Research Centre, for assistance with ß-glucan analyses.

LITERATURE CITED

AASTRUP, S. 1983. Selection and characterization of low ß-glucan mutants from barley. Carlsberg Res. Commun. 48:307-316.
AASTRUP, S. 1988. Application of the Calcofluor flow injection analysis method for determination of ß-glucan in barley, malt, wort, and beer. J. Am. Soc. Brew. Chem. 46:76-81.
AASTRUP, S., and MUNCK, L. 1985. A ß-glucan mutant in barley with thin cell walls. Pages 291-296 in: New Approaches to Research on Cereal Carbohydrates. R. D. Hill and L. Munck, eds. Elsevier Science Publishers, Amsterdam.
AMERICAN ASSOCIATION OF CEREAL CHEMISTS. 1989. AACC committee adopts oat bran definition. Cereal Foods World 34:1033.
BARNES, P. J., and FULCHER, R. G. 1989. Fluorometric measurement of fats. Pages 207-214 in: Fluorescence Analysis of Foods. L. Munck, ed. Longman Scientific and Technical, Longman Group U.K. Ltd , Essex.
BERESFORD, G., and STONE, B. 1983. (1→3),(1→4)-ß-D-glucan content of Triticum grains. J. Cereal Sci. 1:111-114.
COLLINS, F. W. 1986. Oat phenolics: Structure, occurrence, and function.

Pages 227-295 in: Oats: Chemistry and Technology. F. H. Webster, ed. Am. Assoc. Cereal Chem., St. Paul, MN.

COLLINS, F. W. 1989. Oat phenolics: Avenanthramides, novel substituted *N*-cinnamoylanthranilate alkaloids from oat groats and hulls. J. Agric. Food Chem. 37:60-66.

COLLINS, F. W., McLACHLAN, D. C., and BLACKWELL, B. A. 1991. Oat phenolics: Avenalumic acids, a new group of bound phenolic acids from oat groats and hulls. Cereal Chem. 68:184-189.

FINCHER, G. B. 1975. Morphology and chemical composition of barley endosperm cell walls. J. Inst. Brew. 81:116-122.

FINCHER, G. B. 1976. Ferulic acid in barley cell walls: A fluorescence study. J. Inst. Brew. 82:347-349.

FULCHER, R. G. 1972. Observations on the aleurone layer with emphasis on wheat. Ph.D. thesis. Monash University, Melbourne, Australia.

FULCHER, R. G. 1986. Morphological and chemical organization of the oat kernel. Pages 47-74 in: Oats: Chemistry and Technology. F. H. Webster, ed. Am. Assoc. Cereal Chem., St. Paul, MN.

FULCHER, R. G., and WONG, S. I. 1980. Inside cereals—Fluorescence microchemical view. Pages 1-25 in: Cereals for Food and Beverages. G. E. Inglett and L. Munck, eds. Academic Press, New York.

FULCHER, R. G., and WOOD, P. J. 1983. Identification of cereal carbohydrates by fluorescence microscopy. Pages 111-147 in: New Frontiers in Food Microstructure. D. B. Bechtel, ed. Am. Assoc. Cereal Chem., St. Paul, MN.

FULCHER, R. G., O'BRIEN, T. P., and LEE, J. W. 1972a. Studies on the aleurone layer. I. Conventional and fluorescence microscopy of the cell wall with emphasis on phenol-carbohydrate complexes in wheat. Aust. J. Biol. Sci. 25:23-34.

FULCHER, R. G., O'BRIEN, T. P., and SIMMONDS, D. H. 1972b. Localization of arginine-rich proteins in mature seeds of some members of the Gramineae. Aust. J. Biol. Sci. 25:487-497.

FULCHER, R. G., O'BRIEN, T. P., and WONG, S. I. 1981. Microchemical detection of niacin, aromatic amine, and phytin reserves in cereal bran. Cereal Chem. 58:130-135.

FULCHER, R. G., IRVING, D. W., and DE FRANCISCO, A. 1989. Fluorescence microscopy: Applications in food analysis. Pages 59-109 in: Fluorescence Analysis of Foods. L. Munck, ed. Longman Scientific and Technical, Longman Group U.K. Ltd., Essex.

GATES, P. J., and OPARKA, K. J. 1982. The use of the fluorescent probe 9-anilino-1-naphthalene sulphonic acid (ANS) as a histochemical stain in plant tissues. Plant Cell Environ. 5:251-256.

GUILBAULT, G. G. 1989. Principles of fluorescence spectroscopy in the assay of food products. Pages 33-58 in: Fluorescence Analysis of Foods. L. Munck, ed. Longman Scientific and Technical, Longman Group U.K. Ltd., Essex.

HARGIN, K. D., MORRISON, W. R., and FULCHER, R. G. 1980. Triglyceride deposits in the starchy endosperm of wheat. Cereal Chem. 57:320-325.

HAUGLAND, R. 1989. Handbook of Fluorescent Probes and Research Chemicals. Molecular Probes, Inc., Eugene, OR.

HUGHES, J., and McCULLY, M. E. 1975. The use of an optical brightener in the study of plant structure. Stain Technol. 50:319-329.

IRVING, D. W., FULCHER, R. G., BEAN, M. M., and SAUNDERS, R. M. 1989. Differentiation of wheat based on fluorescence, hardness, and protein.

Cereal Chem. 66:471-477.
IRVING, D. W., PEAKE, J. L., and BREDA, V. A. 1991. Nutrient distribution in five perennial grain species exhibited by light and scanning electron microscopy. Cereal Chem. 68:376-382.
JØRGENSEN, K. G. 988. Quantification of high molecular weight (1→3)(1→4)-β-D-glucan using Calcofluor complex formation and flow injection analysis. I. Analytical principle and its standardization. Carlsberg Res. Commun. 53:277-285.
JØRGENSEN, K. G., and AASTRUP, S. 1988a. Quantification of high molecular weight (1→3)(1→4)-β-D-glucan using Calcofluor complex formation and flow injection analysis. II. Determination of total ß-glucan content of barley and malt. Carlsberg Res. Commun. 53:287-296.
JØRGENSEN, K. G., and AASTRUP, S. 1988b. Determination of ß-glucan in barley, malt, wort and beer. Mod. Methods Plant Anal. New Ser. 7:88-108.
McCLEARY, B. V., and GLENNIE-HOLMES, M. 1985. Enzymic quantification of (1→3)(1→4)-β-D-glucan in barley and malt. J. Inst. Brew. 91:285-295.
MILLER, S. S. 1992. Oat ß-glucan: Biochemistry, structure, and genetic variation. Ph.D. thesis. University of Ottawa, Ottawa, Canada.
MILLER, S. S., YIU, S. H., FULCHER, R. G., and ALTOSAAR, I. 1984. Preliminary evaluation of lectins as fluorescent probes of seed structure and composition. Food Microstruct. 3:133-139.
MILLER, S. S., FULCHER R. G., and ALTOSAAR, I. 1989. Evaluation of 4-methylumbelliferyl heptanoate as a substrate for oat lipase. J. Cereal Sci. 10:61-68.
MUNCK, L. 1989. Practical experiences in the development of fluorescence analyses in an applied food research laboratory. Pages 1-32 in: Fluorescence Analysis of Foods. L. Munck, ed. Longman Scientific and Technical, Longman Group U.K. Ltd., Essex.
O'BRIEN, T. P. 1983. Cereal structure, an historical perspective. Pages 3-25 in: New Frontiers in Food Microstructure. D. B. Bechtel, ed. Am. Assoc. Cereal Chem., St. Paul, MN.
O'BRIEN, T. P., and McCULLY, M. E. 1981. The Study of Plant Structure: Principles and Selected Methods. Termacarphi Pty. Ltd, Melbourne, Australia.
PATON, D. 1986. Oat starch: Physical, chemical and structural properties. Pages 93-120 in: Oats: Chemistry and Technology. F. H. Webster, ed. Am. Assoc. Cereal Chem., St. Paul, MN.
PETERSON, D. M., and BRINEGAR, A. C. 1986. Oat storage proteins. Pages 153-203 in: Oats: Chemistry and Technology. F. H. Webster, ed. Am. Assoc. Cereal Chem., St. Paul, MN.
SYMONS, S. J., and FULCHER, R. G. 1988a. Relationship between oat kernel weight and milling yield. J. Cereal Sci. 7:215-217.
SYMONS, S. J., and FULCHER, R. G. 1988b. Determination of variation in oat kernel morphology by digital image analysis. J. Cereal Sci. 7:219-228.
SYMONS, S. J., and FULCHER, R. G. 1988c. Determination of wheat kernel morphological variation by digital image analysis. I. Variation in Eastern Canadian milling quality wheats. J. Cereal Sci. 8:211-218.
SYMONS, S. J., and FULCHER, R. G. 1988d. Determination of wheat kernel morphological variation by digital image analysis: II. Variation in cultivars of soft white winter wheats. J. Cereal Sci. 8:219-229.
WOOD, P. J. 1980. Specificity in the interaction of direct dyes with polysaccharides. Carbohydr. Res. 85:271-287.

WOOD, P. J. 1985. Dye-polysaccharide interactions—Recent research and applications. Pages 267-278 in: New Approaches to Research on Cereal Carbohydrates. R. D. Hill and L. Munck, eds. Elsevier Science Publishers, Amsterdam.

WOOD, P. J., and FULCHER, R. G. 1978. Interaction of some dyes with cereal ß-glucans. Cereal Chem. 55:952-966.

WOOD, P. J., and FULCHER, R. G. 1983. Dye interactions. A basis for specific detection and histochemistry of polysaccharides. J. Histochem. Cytochem. 31:823-826.

WOOD, P. J., FULCHER, R. G., and STONE, B. A. 1983. Studies on the specificity of interaction of cereal cell wall components with Congo Red and Calcofluor. Specific detection and histochemistry of (1→3),(1→4)-ß-D-glucan. J. Cereal Sci. 1:95-110.

WOOD, P. J., WEISZ, J., and BLACKWELL, B. A. 1991a. Molecular characterization of cereal ß-glucans. Structural analysis of oat ß-D-glucan and rapid structural evaluation of ß-D-glucans from different sources by high-performance liquid chromatography of oligosaccharides released by lichenase. Cereal Chem. 68:31-39.

WOOD, P. J., WEISZ, J., and FEDEC, P. 1991b. Potential for (1→3),(1→4)-ß-D-glucan concentrations. Cereal Chem. 68:48-51.

YIU, S. H., POON, H., FULCHER, R. G., and ALTOSAAR, I. 1982. The microscopic structure and chemistry of rapeseed and its products. Food Microstruct. 1:135-143.

YIU, S. H., WEISZ, J., and WOOD, P. J. 1991. Comparison of the effects of microwave and conventional cooking on starch and ß-glucan in rolled oats. Cereal Chem. 68:372-375.

YOUNGS, V. L. 1986. Oat lipids and lipid-related enzymes. Pages 205-226 in: Oats: Chemistry and Technology. F. H. Webster, ed. Am. Assoc. Cereal Chem., St. Paul, MN.

Chapter 2

Processing: Current Practice and Novel Processes

David Paton
Agriculture Canada[1]
Crop Utilization Research Unit
POS Pilot Plant Corporation
Saskatoon, Saskatchewan
S7N 2R4, Canada

Marvin K. Lenz
John Stuart Research Laboratories
The Quaker Oats Company[1]
Barrington, Illinois 60010, USA

Introduction

The nutritional and clinical implications of oat bran as a vehicle for the stabilization of blood glucose levels and a means of lowering serum cholesterol in high-risk subjects are discussed elsewhere, in Chapters 4–6. Since $(1\rightarrow3),(1\rightarrow4)$-β-D-glucan (β-glucan) is the likely active principle, there has been a search for technologies that will yield oat bran fractions enriched in this component. The subaleurone layers of the dehulled oat groat are particularly rich in β-glucan (Chapter 1). Wheat bran, composed mainly of the outer layers of seed coat, pericarp, and aleurone and containing very little starchy endosperm, is prepared by roller milling. Attempts to produce oat bran in ways analogous to those used for wheat bran have met with limited success—a soft kernel characteristic, coupled with a relatively

[1]Mention of specific products or equipment in this article does not constitute an endorsement of these products by Agriculture Canada or The Quaker Oats Co.

high content of lipid in the endosperm, leads to the clogging of rollers and sieves. Commercial oat bran is therefore not an anatomically pure bran but is a sieved fraction that is enriched (up to two times) in β-glucan.

A perusal of the specifications for commercial oat bran made available by 19 processors indicates a wide range for the descriptive characteristics and analytical parameters (Table 1). In addition, commercial oat brans are available in an equally wide range of particle sizes or granulations (Table 2). Only protein and fat content were specified by all processors. Although most provided total dietary fiber (TDF) values, only four companies provided total and soluble dietary fiber (SDF) values as well as a value for β-glucan. Although the variability in the ranges for protein, moisture, fat, and ash is considerable, the primary issue of contention in the media and scientific literature is the contribution of oat bran to the dietary fiber intake of the consumer. With only four of 19 processors reporting TDF and SDF values, a great deal of confusion has existed in the marketplace.

In 1989, in an attempt to address this confusion, the American Association of Cereal Chemists (AACC) adopted a guideline definition for oat bran, as reported in Chapter 1. This definition can only be viewed as an industry guideline, given that a known natural variability exists in the amount of TDF and SDF within oat cultivars grown commercially for use by the food industry in North America and around the world.

TABLE 1
Ranges of Specifications for Oat Bran from Various Commercial Sources

Fraction	Specification
Foreign matter	No hulls, some hulls, 10 hulls/100 g
Enzyme activity	Low to none
Protein content, % dmb[a]	10–21.2
Moisture content, %	8–12.5 (max)
Ash, % dmb	1.1–5.6 (max)
Fat, % dmb	4.4–11.1 (max)
Crude fiber, %	1.1–4.4 (max)
Total dietary fiber, % dmb	12.5–24.0
Soluble dietary fiber, % dmb	5.6–9.4
β-Glucan, % dmb	3.5–8.3

[a]Dry matter basis.

TABLE 2
Ranges of Granulation Sizes for Oat Brans from Various Sources

Screen Size	Percent Granulation
On U.S. no. 14 (1.40 mm)	1–15
On U.S. no. 20 (850 μm)	30–75
On U.S. no. 30 (600 μm)	5–40
On U.S. no. 40 (425 μm)	10–65
Through U.S. no. 40 (425 μm)	1–20

Before discussion of the variations in currently used dry-milling practices to obtain oat bran, some historical perspective is required. The first commercial oat bran was produced by the Quaker Oats Co. in the late 1970s. Oat groats were steamed, flaked, and ground in a hammer mill (Fig. 1). The milled material was sifted on a plansifter using a 36-mesh screen (538 μm) to give a coarse fraction (oat bran) comprising 40–50% of the starting material. The proximate composition of the starting groats and oat bran found by Gould et al (1980) is presented in Table 3. The oat bran fraction was enriched in protein, fat, crude fiber, ash, and dietary fiber. The β-glucan content was twice that found in the starting groat. (The procedure illustrated in Figure 1 is a particle size reduction with a screening step to remove coarse particles. Further separation of endospermic flour from the outer aleurone and subaleurone layers of the oat groat is not possible with this process.)

Current Practice

With some limitations, the roller-milling methods used to make wheat and corn bran can be used to produce oat bran. Such a process was described by Wu et al (1972), together with an analysis of the oat bran obtained by employing a 25-mesh (840-μm) screen on the plansifter. The yield of oat bran was 18.6–25.0%, which was 50% lower

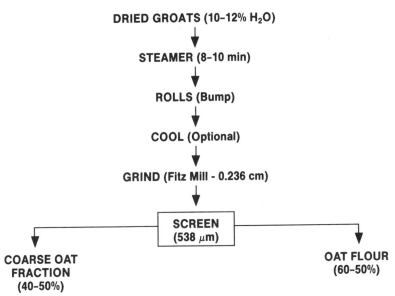

Figure 1. Schematic flow of production of oat bran. (Reprinted, with permission, from Gould et al, 1980)

than that obtained by Gould et al in 1980. The bran protein contents were not substantially different in each of these cases. This is likely more attributable to factors such as the higher starting protein content and perhaps the microstructure of the oat groats than to the process itself.

Some consideration of the current oat-milling practice and how it can affect the quality of the final oat bran is worthwhile. Oat bran production remains essentially a dry grinding and sifting operation. Special attention must be given to oat groat quality before milling, since any foreign matter such as oat hulls, weed seeds, and other grains, which are more vitreous than oat groats and may resist grinding, will end up in the coarse oat bran fraction. Removal of extraneous matter can be achieved using various approaches such as specific gravity tables, plansifters, disk separators, and aspirators to reduce contamination levels. Grinding and sifting alone do not stabilize the milled product streams with respect to enzyme activity—this stabilization must be accomplished in the starting groat. The general oat-milling process begins with a cleaning of the incoming seeds to remove foreign matter. The hull is then detached from the groat by an impact dehuller, and aspiration is used to remove the hull from the groat. Any groats that have not been dehulled are separated by some type of grading device and returned to the impact dehuller. The groats are then kilned to inactivate enzymes and develop flavor. Groats may then be steamed and flaked as whole groats or cut into pieces and steamed and flaked. Either material is suitable for oat bran production.

The presence of an active lipase is particularly troublesome in oat products (Youngs, 1986). Groats contain 6.0–9.0% fat, which is higher than levels found in other commercial cereal grains (Deane and Commers, 1986). Lipases catalyze the hydrolysis of triglycerides to free fatty acids, which cause the oat products to taste rancid and be unpalat-

TABLE 3
Proximate Analysis (% as is) of Coarse Fraction (Oat Bran)
Compared with Starting Groats[a]

	Groat	Coarse Fraction
Protein (N × 6.25)	16.5	21.22
Fat (ether extract)	6.2	7.5
Moisture	9.3	8.5
Crude fiber	1.6	2.7
Ash	1.9	2.6
Neutral detergent fiber	6–9	14–30
β-Glucan	4–6	8–12

[a]Source: Gould et al (1980); used by permission.

able. Whole oats or cut oats that have been kilned still retain 20–40% of their original lipase activity. The remaining lipase is inactivated by steaming before the groat is rolled into flakes. Oat flakes or kilned groats may be used to produce commercial oat bran.

Figure 2 is a general flow sheet for producing oat bran from two options, namely the hammer mill and roller mill methods. While either approach is capable of yielding oat brans that meet the AACC definition, little published information is available on the relative merits of each. Roller milling of oats is generally acknowledged to be inefficient because of clogging of roller serrations and screens by the soft endospermic flour. Variation in granulation of the oat bran is made possible by controlling the speed and screen openings in hammer milling and the roller gap and screen openings in roller milling. Granulation is an important contributor to the final sensory characteristics of oat bran products. The final oat bran moisture content is usually 1–2% lower than the moisture content of the groat or flakes used as the starting material. Final moisture levels are preferentially kept below 12% to restrict the growth of molds during storage (Todd, 1980). Change of screen sizes in any of the sifters results in variations in bran fraction yield and composition; higher bran fraction yields give rise to lower amounts of bran β-glucan and TDF (Wood et al, 1989). Typical screen sizes are 25 mesh (710 μm) and 36 mesh (538 μm), but a 48-mesh (390-μm) size can also be used (J. Schroeder, *personal communication*).

Following removal of the coarse bran fraction, the material passing

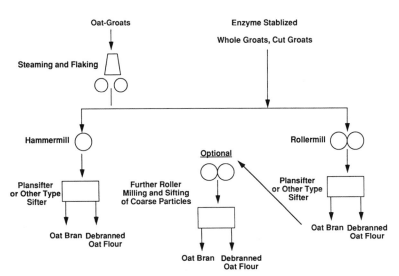

Figure 2. Generalized flow of production of oat bran showing possible processing options.

through the screens or sifters is recovered as a partially debranned oat flour, the granulation of which is close to that of commercial whole oat flour. While there are no published values comparing partially debranned oat flour with whole oat flour, a mass balance would indicate that: the TDF would be lowered from 10.5% in whole oat flour to 8.0–9.3% in partially debranned flour; the protein content would be lowered from 17.0% to 14.5–16.0%, and the β-glucan content lowered from 4.5% to 3.4–4.2%. While in many applications, partially debranned oat flour and whole oat flour can be interchanged without loss of functionality, in some cases, e.g., in extrusion cooking, a low β-glucan oat flour is preferred and may offer specific functional advantages such as increased product expansion (Gordon et al, 1986).

As human nutritional studies on the role of oat bran and β-glucan continue, the importance of oat bran fractions more highly enriched in β-glucan will probably increase. The value of such material will relate to the amount of the active ingredient or carrier of the ingredient that can be practically incorporated into formulated foods such as bread, cakes, and muffins. The remainder of this chapter discusses experimental processes that have been examined at the pilot-plant level and that produce oat bran fractions enriched in β-glucan. Some functional properties of these are compared with those of other fiber sources, including some that are low in or devoid of an SDF component.

In the early 1970s, the main thrust in cereal, pulse, and oilseed processing was to produce fractions enriched in protein. In one such study, Wu and Stringfellow (1973) examined two oat cultivars, Sioux and Garland. Dehulled seed was pin milled (three times at 14,000 rpm) and defatted with pentane-hexane to give materials of 16.3 and 22.7% protein, respectively. Each flour was subjected to air classification. The system was set up such that each pass of material through the classifier gave a coarse and a fine fraction. By adjusting the cut points to 15, 18, 24, and 30 μm in successive passes, four fine fractions (1B, 2, 3, and 4) and a coarse fraction (5) were obtained. In addition, an ultrafine fraction, 1A, was collected in the air filter bag during the first pass through the classifier. The results are shown in Table 4. Since the focus was on protein-enriched flours, the bran fraction was considered a waste by-product. However, this fraction was three times enriched in crude fiber content. One might extrapolate that the TDF content would be high. The yield (22–24%, w/w) might also suggest that this fraction was relatively free of inner endosperm and was enriched in β-glucan.

Sosulski and Sosulski (1985), in cooperation with a commercial oat mill, devised a procedure for separating wild oats (*Avena fatua* L.)

from all other seeds contained in screenings from a seed-cleaning operation and dehulled them as a first step in preparing a lipase-inactivated wild oat flake. To prepare a coarse bran and flour fraction, the groats were milled on an Allis-Chalmers mill using two sets of corrugated rolls and three settings of gap size. Following each pass, the fractions collected on 30GG (grit gauge 727 μm) and 60GG (308 μm) metal screens were combined and designated as bran. The yield was 40% (w/w) for wild oat groat compared to 35% (w/w) for a common oat (*A. sativa*). Continuous roller milling is limited by the slowness of the sieving steps. Wild oat groats were shown to contain higher levels of protein, P, Cu, Fe, Zn, vitamin A, riboflavin, and thiamin than the common oat.

A process patent issued to the Quaker Oats Company (Hohner and Hyldon, 1977) described, in part, a dry-milling and air-classification procedure for the preparation of a bran fraction from hexane-defatted oat flake. Two oat samples were used in this study: Dal, a high-protein high-lipid variety, and regular oats, presumably as supplied through the usual purchasing channels that supply non-variety-specific grain. No details were given on either groat or coarse bran composition or on the yields of crude oil from the original oat flakes. The yield of air-classified coarse-bran fraction from the cultivar Dal was 27.7%. The coarse fraction identified in Figure 3 was then subjected to

TABLE 4
Analysis of Air-Classified Fractions of Ground
Sioux and Garland Oat Groats[a,b]

Fraction	Yield (%)	Protein Percent	Protein Percent of Total Groat Protein	Fat (%)	Fiber (%)
Sioux groats					
Ground[c]	...	16.3	...	1.4	1.6
1A	2	83.3	10.2	...	...
1B	26	20.8	33.2	1.3	0.1
2	16	13.1	12.8	1.1	0.1
3	25	9.0	13.8	0.9	0.9
4	9	8.4	4.6	0.9	1.6
5	22	24.2	32.7	3.2	5.0
Garland groats					
Ground[c]	...	22.7	...	1.4	2.2
1A	2	83.1	7.3	...	...
1B	26	29.4	33.7	1.3	0.0
2	16	18.7	13.2	1.2	0.0
3	24	12.3	13.0	1.1	0.9
4	8	11.8	4.2	1.2	2.0
5	24	29.2	30.9	2.8	6.4

[a]Source: Wu and Stringfellow (1973); used by permission.
[b]Expressed on a dry matter basis.
[c]Three times at 14,000 rpm.

an aqueous extraction in sodium carbonate solution at pH 9.2, which solubilized most of the protein and the β-glucan (Fig. 4).

The slurry was centrifuged at low speed to remove the extracted bran, leaving the oat starch suspended in the protein plus β-glucan liquor. Following precipitation of protein at the isoelectric point and its removal with the starch as a coproduct, the remaining liquor was neutralized, concentrated, and precipitated with 2-propanol (isopropyl alcohol) to yield an oat gum. Although no compositional information

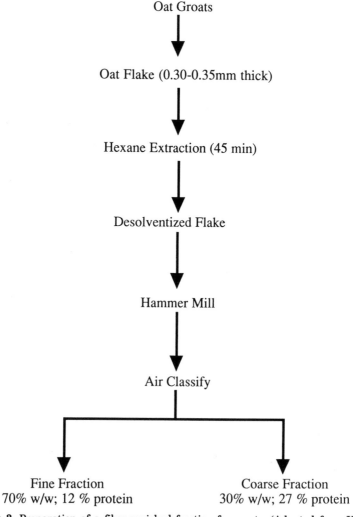

Oat Groats

Oat Flake (0.30-0.35mm thick)

Hexane Extraction (45 min)

Desolventized Flake

Hammer Mill

Air Classify

Fine Fraction   Coarse Fraction
70% w/w; 12 % protein   30% w/w; 27 % protein

Figure 3. Preparation of a fiber-enriched fraction from oats. (Adapted from Hohner and Hyldon, 1977)

was given for the coarse bran fraction, it is reasonable to assume that this fraction was somewhat enriched in β-glucan as a result of the procedures employed. Further, no data was given for the purity of the β-glucan-containing oat gum. However, studies elsewhere (Chen et al, 1981) reported that an oat gum obtained from the Quaker Oats Company contained 66% β-glucan.

Novel Processes

AQUEOUS PROCESSING

Following up on observations made during fundamental studies of seed dormancy, Burrows et al (1984) developed a process whereby whole oats or cleaned oat groats were steeped in water for up to 28 hr at 50°C. The steeped kernels were extremely turgid, and it was found that the endosperm cell walls had been sufficiently disrupted to allow the endosperm contents to be easily removed from the outer hull and the aleurone and subaleurone layers.

During the steeping process, the enzymes of the subaleurone layers and embryo tissues evidently remain compartmentalized or are inactive, since hydrolysis of starch, protein, and lipid did not occur within the turgid kernel. A flow diagram for the process is illus-

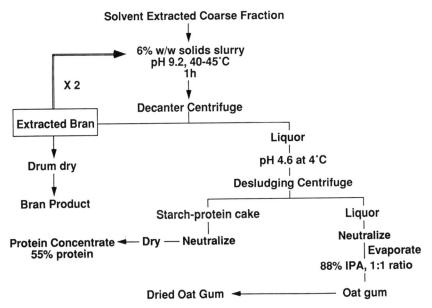

Figure 4. Preparation of oat gum and oat protein concentrate from a fiber-enriched fraction from oats. IPA = isopropyl alcohol. (Adapted from Hohner and Hyldon, 1977)

trated in Figure 5. In the original patent specification, the liquid medium used to grind the seed was water. Although no starch or protein hydrolysis occurred upon wet milling, oat lipids were hydrolyzed as a result of grain disruption and the oat lipase coming into contact with its substrate.

The bran and flour fractions produced were of unacceptable quality due to high levels of objectionable free fatty acids. An improvement to this process was filed wherein the liquid medium used to macerate the steeped seeds was a water-miscible aliphatic alcohol (Collins and Paton, 1990). Use of an alcohol was found to have two benefits: prevention of lipase action and a more facile liquid medium from which bran and endosperm may be fractionated. The primary bran produced is a stabilized flake and is relatively free of adhering starchy endosperm.

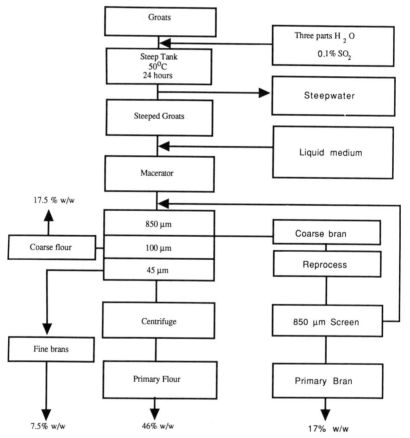

Figure 5. Processing of oat groats to produce fiber-enriched and fiber-depleted fractions. (Adapted from Burrows et al, 1984)

Ten replicate steeps, each of 4 kg (dmb) of naked oats (cv. Tibor), were processed using this improved procedure. The yield of primary bran was $15.07 \pm 0.44\%$ (w/w) with the following composition of dry matter: protein (N × 6.25), $23.3 \pm 0.2\%$; hexane-extractable fat, $10.6 \pm 0.42\%$; starch, $17.5 \pm 0.3\%$; and β-glucan, $14.7 \pm 0.2\%$. The β-glucan content represents a three-fold enrichment over the content of the groat (4.7%).

A recent publication by Wilhelm et al (1989) describes a wet-milling process for oats wherein the groats are dry milled and then the comminuted groats are soaked in a solution of cellulolytic and hemicellulolytic enzymes. The process is illustrated in Figure 6. After the digest was screened to remove residual fibrous material, the liquid phase was subjected to a multihydrocyclone classification, producing

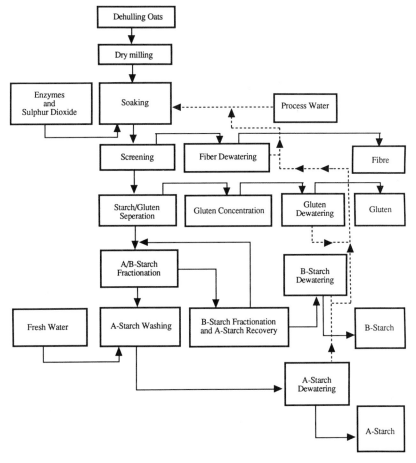

Figure 6. Preparation of starch and protein from an enzymatic digest of milled oat groats. (Reprinted, with permission, from Wilhelm et al, 1989)

a protein-enriched flour (50–55% protein), a prime starch fraction (1% protein), and a secondary starch fraction (10–15% protein). The bran had a protein content of 6%. This study focused on starch production: 55–65% of the total starch was found in the A-starch product, 21–31% in the B-starch, 10% in the protein fraction, and 2% in each of the fiber and process water streams. The residual bran from this process is highly degraded and likely contains very little β-glucan.

Inglett (1992) described a process for the preparation of an oat SDF fraction. Although not directly related to oat bran, this process nevertheless has some relevance to the subject matter since the final product contains at least the same β-glucan level as most commercial oat brans. The process is described in Figure 7.

The enzyme used was a thermostable α-amylase having the essential characteristics of amylases produced by strains of *Bacillus stearothermophilus*, genetically modified *B. subtilis*, or *B. licheniformis*. These enzymes are available commercially under the names Enzeco Thermolase (Biddle Sawyer Corp., New York) and Takalite (Miles Laboratories, Elkhart, IN). Starch was converted into maltodextrins with preferred degrees of polymerization of 6 (13%) and 5 (14%); some of

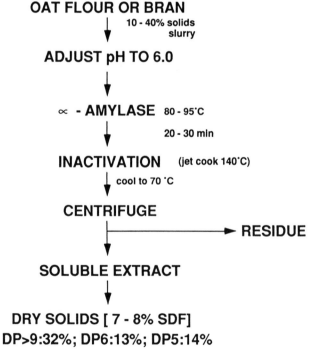

OAT FLOUR OR BRAN
10 - 40% solids
slurry

ADJUST pH TO 6.0

∝ - AMYLASE 80 - 95°C
20 - 30 min

INACTIVATION (jet cook 140°C)
cool to 70 °C

CENTRIFUGE ——————→ **RESIDUE**

SOLUBLE EXTRACT

DRY SOLIDS [7 - 8% SDF]
DP>9:32%; DP6:13%; DP5:14%

Figure 7. Process flow sheet for the production of Oatrim. SDF = soluble dietary fiber, DP = degree of polymerization. (Adapted from Inglett, 1990)

the oat β-glucan was co-solubilized. Following enzyme inactivation, the insoluble residues were removed by centrifugation, and the extract was dried to give a white powder known by the trade name Oatrim. The product was found to contain 7–8% SDF (oat β-glucan) and is reported to have uses as a fat substitute in products such as ice creams.

Lehtomaki et al (1990) homogenized oat groats in cold water (<10°C) and then wet-screened them successively on 800- and 80-μm sieves. The fraction retained on the 80-μm sieve was washed with cold water and then rapidly dried. Although very few process details were given, the β-glucan of the bran preparation was claimed to be in the range of 15–30% (w/w). Literature supplied by the assignee (Alko AB, Finland) lists the β-glucan content of the bran product at 14.0% and the TDF at 35.0%.

NONAQUEOUS PROCESSING

Oughton (1980a,b) produced fractions of high protein content by fractionating slurries of oats in nonaqueous organic solvents. Using a plurality of liquid cyclones and screening techniques, oats were fractionated into high- and low-protein streams, bran, and oil. A typical example is illustrated in Figure 8. Products 1, 4, and 6 are flour fractions of varying protein contents; products 2, 3, and 5 are brans of essentially similar protein content. The total yield of brans was 28.5%. Although the brans were likely enriched in β-glucan, this component was not quantified. Attempts to measure viscosity using a Brookfield model RVT viscometer, fitted with a no. 1 or 2 spindle at 60 rpm, were unsuccessful because of the high, pastelike viscosities produced in an aqueous medium at the concentration chosen. Bran was not the prime target of these experiments. The various combinations of techniques produced yields of coarse fractions in the range of 28–50% (w/w). Recognizing the preference for using economical and available equipment, Boczewski (1980) described a further development of Oughton's work in which oat groats were passed between a selected number of paired smooth rolls, the crushed oats mixed with an organic solvent (hexane), and the slurry screened and processed using hydrocyclones to produce a number of fractions of varying protein content. Each pair of rollers was capable of being operated at differential rotational speeds under variable spring-loaded tension; the roller gap was also adjustable. Using the oat cultivar Hinoat, the yield of bran recovered ranged from 26 to 53% (w/w), depending upon the number of roller pairs (or passes through one roller pair) used. A slurry of the flour fraction in hexane was fed to a 10-mm hydrocyclone having a 12-mm long vortex finder and a diameter of 6.75 mm.

At an operating pressure of 689 kPa, the overflow yielded a flour of 9% (dmb) at a protein content of 85%, whereas the underflow (91% yield) had a protein content of 10%. These methods gave results similar to those of Oughton (1980a,b) but with the advantage of a simpler front-end comminution. The studies of Oughton (1980a,b) and Boczewski (1980) did not consider bran as a primary product.

NONAQUEOUS PLUS AQUEOUS PROCESSING

In the mid 1980s, the human physiological significance of oat bran and its β-glucan component became more evident; oat bran was no longer considered a by-product of oat processing but was a sought-after commodity. Oat bran enriched in β-glucan is now a valued material. To study the physiological action of oat β-glucan, Wood et al (1989) prepared large quantities of oat bran, which could be used to extract β-glucan in greater yields than possible from whole groats. A flow chart (Fig. 9) shows three different processes for producing bran. The overall yield of the hexane-extracted, air-classified bran

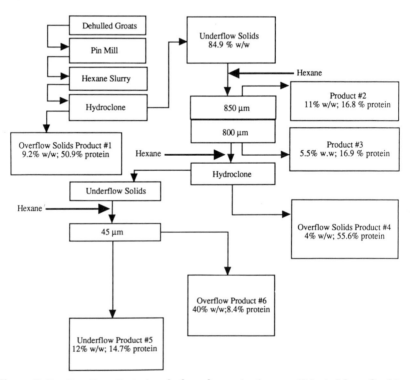

Figure 8. Fractionation of oats in a hydrocyclone using hexane. (Adapted from Oughton, 1980a)

fraction was 38% (w/w), having a β-glucan content of 11.2 ± 0.3%. Following refluxing in 75% ethanol to inactivate enzymes, the gum was extracted from 200 kg of the fraction, essentially using the methods of Hohner and Hyldon (1977) (Fig. 10, process A). The yield of gum obtained was 18.6 kg, which represented a β-glucan extraction of 66.0%. Further studies were conducted to evaluate the feasibility of air classification of full-fat flour and to obtain bran of higher β-glucan content. These are represented by processes B and C in Figure 10. The β-glucan contents were 12.8% and 16.6%, respectively. The data of Wood et al (1989) and F. W. Collins and D. Paton (*unpublished data*) thus demonstrate that it is possible to obtain oat bran fractions containing a substantially higher level of β-glucan (14.7–19.1%) than in any of the fractions presently commercially available.

Myllymaki et al (1989) also described a process for obtaining oat bran enriched in β-glucan (Fig. 11). This represents only a portion of a process intended to fractionate oats into industrial raw materials.

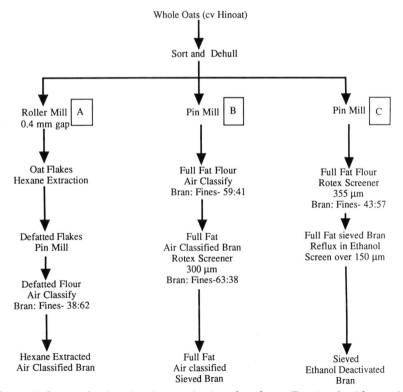

Figure 9. Options for the pilot-plant production of oat bran. (Reprinted, with permission, from Wood et al, 1989)

The process appears to incorporate elements of other technologies (Hohner and Hyldon, 1977; Oughton, 1980a,b) to prepare starch, protein, and β-glucan. The step used to obtain β-glucan-enriched oat bran is similar to that published by Wood et al (1989).

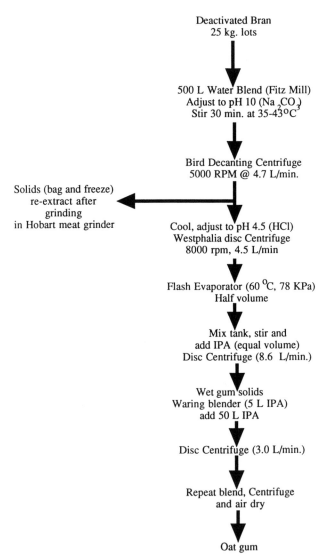

Figure 10. Preparation of pilot-plant quantity of oat gum. IPA = isopropyl alcohol. (Reprinted, with permission, from Wood et al, 1989)

Properties of Fiber-Rich Oat Brans

Most dietary fiber preparations are incorporated into foods (bread, muffins, cookies, crackers, breakfast cereals, etc.) either as a bulking agent or as a granulated topping. An aim of such food products is often to incorporate as much of the fiber ingredient as possible without causing significant negative effects to product characteristics such as volume, texture, or mouthfeel. Few studies have been published that indicate that such bran materials might possess positive functional properties rather than negative characteristics that restrict use. Chang and Sosulski (1985) compared functionality of roller-milled fractions from common and wild oats (Table 5). All of the oat fractions (whole meal, bran, and flour) exhibited a higher water absorption capacity than wheat, corn, or rice flours, with oat bran having the highest. Oat brans were also higher in oil absorption and emulsion capacity than other oat fractions or the reference flours. The substantially higher values for the apparent viscosity of cold-water

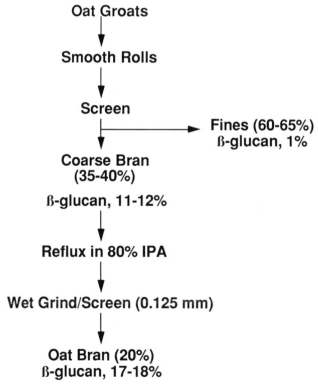

Figure 11. Preparation of β-glucan-enriched oat bran. IPA = isopropyl alcohol. (Adapted from Myllymaki et al, 1989)

bran suspensions presumably reflects the enrichment of β-glucan in these brans, although β-glucan content was not reported. The high values for all functional characteristics of the potato flour are presumably due to starch, since commercial potato flours are produced using a roller-drying technique that cooks the starch component. The water hydration capacity (WHC) of fiber-rich materials is perhaps the most important functional test, as it has a marked bearing on dough mixing when such ingredients are incorporated into bakery products. Table 6 shows the WHCs of a range of commercially available high-fiber ingredients and two experimental fiber-enriched oat brans. All samples were ground to pass a 0.177-mm (U.S. standard 80-mesh) screen before the test. The data clearly show the differences among the highly lignified fiber sources (wheat and rice brans and oat hull fiber) and the others. Apple and sugar beet fibers are different, these being predominantly the dried cell wall residues after juice extraction. Whereas the juices contain some colloidal pectin, the residual cell wall material contains an insoluble form of pectin, protopectin, in addition to cellulose and hemicellulose. In commercial pectin production from dried apple, citrus, or sugar beet residues, strong mineral acid is required to solubilize the protopectin (Kertesz, 1951). (It is therefore unlikely that the protopectin contributes substantially to WHC at room temperature.) More likely, the higher WHC value for apple and beet fibers is a function of the dehydration

TABLE 5
Some Functional Properties of Milled Fractions Derived from Oats[a]

Crop and Milled Fraction	pH of Dispersion	Water Hydration Capacity (g/g)	Oil Absorption (%)	Apparent Viscosity[b] (cP)	Emulsion Capacity (g/g)
Domestic groats					
Meal	6.4	1.18	5	7.8	29.0
Bran	6.5	2.17	137	42.0	35.3
Flour	6.2	1.10	62	6.8	28.5
Wild groats					
Meal	6.3	1.30	73	8.8	29.1
Bran	6.4	1.91	120	29.8	33.7
Flour	6.1	1.20	56	6.0	27.3
Defatted wild groats					
Meal	6.3	1.32	88	9.3	30.9
Bran	6.4	2.29	178	50.8	34.8
Flour	6.2	1.24	80	7.8	29.4
Wheat flour	5.6	0.93	101	2.1	29.4
Corn flour	6.0	0.92	73	2.9	15.6
Rice flour	6.2	0.86	75	2.3	13.8
Potato flour	6.0	5.23	129	140.0	26.6

[a]Source: Chang and Sosulski (1985); used by permission.
[b]Viscosity of 10% solids slurry at room temperature, using a Haake RV-3 viscometer equipped with an NV sensor and operated at 90.6 rpm.

conditions employed. Careful drying may preserve some of the capillary cellular structure, thereby promoting higher water uptake (Paton, 1974).

The β-glucan contents of the commercial and the experimental A and B oat brans were 8.0, 16.6, and 19.1%, respectively. In AACC Method 88-04 (AACC, 1983), water is incrementally added to the solid with stirring and centrifugation until the point is reached at which excess water is just apparent on the surface of the centrifuged sample. The WHC of oat brans A and B could only be estimated. When water is added initially to these oat bran powders, absorption occurs; further additions of water result in dispersion, partial solubilization, and development of viscosity. The values for estimated WHC represent the absorptive stage only. The behavior of dispersion, partial solubilization, and viscosity development may be attributed to the high β-glucan content and the ready availability of this component for hydration and solubilization. This characteristic is a function of the process employed in the preparation of these oat brans. None of the other fiber products exhibited this characteristic. The relationship between viscosity and shear rate for aqueous suspensions of high-β-glucan oat bran fractions is shown in Figure 12. The β-glucan content of the brans was 19.1% (bran 1), 15.7% (bran 2), and 14.0% (bran 3). Brans 1 and 2 were prepared by the process of Collins and Paton (1990), whereas Bran 3 was prepared according to the procedure of Wood et al (1989). Each bran was ground in a coffee mill, and the entire sample was sieved through a 0.177-mm (80-mesh) screen. One gram (dmb) of each bran powder was wetted with 1 ml of 95% ethanol followed by distilled water to a total weight of 20 g. The pH of each

TABLE 6
Water Hydration Capacity of Selected Plant Fibers

Fiber Source	Water Hydration Capacity[a] (g/g)
Wheat bran	1.67 ± 0.13
Sugar beet	5.90 ± 0.15
Lupine hull	3.40 ± 0.10
Rice bran	1.61 ± 0.18
Apple fiber	3.24 ± 0.13
Barley bran	2.33 ± 0.19
Pea fiber	2.81 ± 0.18
Commercial oat bran	4.12 ± 0.23
Oat hull fiber	1.76 ± 0.10
Experimental oat bran 3 (Wood et al, 1989)	7.76[b] ± 0.26
Experimental oat bran 2 (Collins and Paton, 1990)	12.74[b] ± 0.28

[a] By AACC method 88-04.
[b] Samples begin to flow. No excess water visible.

5% solids slurry was adjusted to 6.5 by the addition of $0.5N$ sodium hydroxide solution. Samples were rapidly stirred until uniformly slurried and then were left for 60 min at room temperature (22°C) with intermittent stirring (brans 1 and 2). In the case of bran 3, insufficient viscosity was developed within 60 min to maintain the residual bran particles in suspension. A total development time of 3.5 hr was required to avoid sedimentation. Seven grams of each slurry was placed in the cup of an NV-1 sensor system of a Haake Rotoviscometer (model RV-100). An M500 measuring head was used, and the cup was thermostated at 20 ± 0.1°C. Data points were collected over 12 min, processed by the computer software, and fitted to a log-log plot based on a power law model:

$$\tau \text{ (shear stress)} = K \gamma^n \text{ (shear rate)} .$$

Table 7 gives the regression coefficients, consistency index (K) and flow index (n) for each bran. The index K is a computed extrapolation and represents a theoretical apparent viscosity (Pa.sec) at 1 sec^{-1}. The higher K for Marion bran (bran 1) presumably reflects the higher β-glucan content when compared to Hinoat (bran 2), since the two brans were identically processed and analyzed at the same solids level (5%, w/w). Adjusting the solids level of bran 2 to give the same

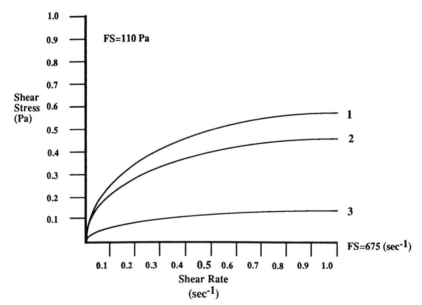

Figure 12. Flow behavior of 5% w/w dispersions of oat bran powder. FS = full scale. (D. Paton, *unpublished data*)

β-glucan level in the slurry as that of bran 1 results in similar values for K and n. On the other hand, bran 3 produced a lower viscosity dispersion with a lower value for K. Increasing the solids level to achieve similar β-glucan content in the slurries did not similarly raise viscosity to the level of bran 2. These results indicate that the methods used to produce oat bran from groats influence the functionality even though the β-glucan content may be identical.

The viscosity-building potential of high β-glucan oat bran dispersions can be further enhanced by heat. Cooling the heated dispersion results in a semisolid mass that exhibits a yield point. Table 8 shows the flow behavior of a 5% dispersion of bran 2 heated to 60°C but cooled to 20°C before analysis. The dispersion was first sheared in the viscometer at 675 sec^{-1} to destroy the yield point; the flow curve was then determined over the shear rate range 0–675 sec^{-1}. The value of K at time zero was 8.75 Pa.sec at 1 sec^{-1} compared with a value of 2.9 Pa.sec for the unheated bran 2 dispersion shown in Table 7. Although the values for K are different, the flow indices are similar. Table 8 also shows that the viscosity of bran 2 dispersions, stored over a five-day period, was unchanged, indicating the absence of β-glucanase activity.

The experimental oat brans described here exceed the specified minimums of the recent definition adopted by the AACC. For

TABLE 7
Power Law Constants for 5% Dispersions of Oat Bran Powders

Bran Source[a]	Consistency Index, K (Pa.sec^{-1})	Flow Behavior Index, n	Correlation Coefficient, r^2
Bran 1	7.0 ± 0.25	0.34 ± 0.02	0.998
Bran 2	2.9 ± 0.11	0.42 ± 0.03	0.998
Bran 3	0.5 ± 0.08	0.55 ± 0.06	0.973
Bran 2[b]	7.1 ± 0.23	0.35 ± 0.02	0.999

[a]Bran identification same as shown in Fig. 12.
[b]Solids level increased to obtain the same β-glucan content as Bran 1.

TABLE 8
Effect of Storage at 5°C on the Flow Behavior of 5% (w/w) Dispersions of β-Glucan-Enriched Oat Bran[a,b]

Time (days)	Consistency Index, K (Pa.sec^{-1})	Flow Behavior Index, n
0	8.57	0.442
1	8.68	0.446
2	8.52	0.440
5	8.65	0.448

[a]Bran 2 in text.
[b]Stirred at 60°C (5 min), cooled, and stored at 5°C. After storage, sample warmed to 20°C, sheared at 675 sec^{-1} (5 min) in viscometer to destroy yield point before flow behavior was recorded.

example, a bran prepared as described by Collins and Paton (1990) was found to have a TDF of 49.9%, comprised of 19.2% SDF and 30.7% insoluble dietary fiber. The minimums adopted by AACC suggest 16.0% TDF, not less than one third of which is SDF. The experimental brans were much enriched in both the soluble and insoluble dietary fiber components. However, as bran purity is increased, the yield is decreased and more debranned flour is produced. This puts a strain on companies to identify new uses for oat flour. This problem may be compounded with the implementation of novel processes that produce high-purity oat brans. Research into novel utilization of oat flour, however, may well lead to improved added value. Such findings may make these experimental processes commercially attractive in the last decade of the 20th century and greatly assist in ensuring the viability of the oat industry into and beyond the year 2000.

ACKNOWLEDGMENT

This material represents Contribution 1023 from the Saskatoon Research Station.

LITERATURE CITED

AMERICAN ASSOCIATION OF CEREAL CHEMISTS. 1983. Approved Methods of the AACC, 8th ed. Method 88-04, approved September 1978, reviewed October 1982. The Association, St. Paul, MN.

BOCZEWSKI, M. P. 1980. Process for the treatment of oats. U.S. patent 4,220,287.

BURROWS, V. D., FULCHER, R. G., and PATON, D. 1984. Processing aqueous treated cereals. U.S. patent 4,435,429.

CHANG, P. R., and SOSULSKI, F. W. 1985. Functional properties of dry milled fractions from wild oats (Avena fatua L.). J. Food Sci. 50:1143-1147.

CHEN, W.-J. L., ANDERSON, J. W., and GOULD, M. R. 1981. Effects of oat bran, oat gum and pectin on lipid metabolism of cholesterol-fed rats. Nutr. Rep. Int. 24:1093-1097.

COLLINS, F. W., and PATON, D. 1990. Recovery of added values from cereal wastes. Canadian patent appl. 2,013,190.

DEANE, D., and COMMERS, E. 1986. Oat cleaning and processing. Pages 371-412 in: Oats: Chemistry and Technology. F. H. Webster, ed. Am. Assoc. Cereal Chem., St. Paul, MN.

GORDON, W. A., HEMPENIUS, W. L., and KIRKWOOD, J. R. 1986. Process for preparing a highly expanded oat cereal product. U.S. patent 4,620,981.

GOULD, M. R., ANDERSON, J. W., and O'MAHONEY, S. 1980. Biofunctional properties of oats. Pages 447-460 in: Cereals for Food and Beverages. G. E. Inglett and L. Munck, eds. Academic Press, New York.

HOHNER, G. A., and HYLDON, R. G. 1977 Oat groat fractionation process. U.S. patent 4,028,468.

INGLETT, G. E. 1992. Oat soluble dietary fiber compositions. U.S. patent 5082673.

KERTESZ, Z. I. 1951. The pectic substances. Interscience Publishers, New York.

LEHTOMAKI, I., KARINEN, P., BERGELIN, R., and MYLLYMAKE, O. 1990. A β-glucan enriched alimentary fiber and process for preparing the same. EP application A2/0377530.

MYLLYMAKI, O., MALKI, Y., and AUTIO, K. 1989. A process for fractionating crops into industrial raw materials. PCT application W089/01294.

OUGHTON, R. W. 1980a. Process for the treatment of comminuted oats. U.S. patent 4,211,695.

OUGHTON, R. W. 1980b. Process for the treatment of comminuted oats U.S. patent 4,211,801.

PATON, D. 1974. Isolation, purification, chemical modification and properties of alcohol insoluble solids (AIS) from the edible portion of apple tissue. Can. Inst. Food Sci. Technol. J. 7(1):65-67.

SOSULSKI, F. W., and SOSULSKI, K. 1985 Processing and composition of wild oat groats (Avena fatua L.) J. Food. Eng. 4:189-203.

TODD, M. E. 1980. Psychrometrics applied to grain groccesing. An updated analysis and application. Publication 13-81:3-5. American Society of Agricultural Engineers, St. Joseph, MI.

WILHELM, E., KEMP, F. W., LEHMUSSAARI, A., and CARANSA, A. 1989. Verfahren zur Herstellung von starke, protein und Fasern ans Hafer. Starch/Staerke 41:372-376.

WOOD, P. J., WEISZ, J., FEDEC, P., and BURROWS, V. D. 1989. Large-scale preparation and properties of oat fractions enriched in (1→3)(1→4)-β-D-glucan. Cereal Chem. 66:97-103.

YOUNGS, V. L. 1986. Oat lipids and lipid-related enzymes. Pages 205-226 in: Oats: Chemistry and Technology. F. H. Webster, ed. Am. Assoc. Cereal Chem., St. Paul, MN.

WU, Y. V., and STRINGFELLOW, A. C. 1973. Protein concentrates from oat flours by air classification of normal and high-protein varieties. Cereal Chem. 50:489-496.

WU, Y. V., SEXSON, K. R., CAVINS, J. F., and INGLETT, G. E. 1972. Oats and their dry-milled fractions. Protein isolation and properties of four varieties. J. Agric. Food Chem. 20:757-761.

Chapter 3

Comparisons of Dietary Fiber and Selected Nutrient Compositions of Oat and Other Grain Fractions

Judith A. Marlett
Department of Nutritional Sciences
University of Wisconsin-Madison
Madison, Wisconsin 53706, USA

Introduction

For many years oats were consumed in the form of hot oatmeal porridge or as an ingredient in baked goods. Their chief nutritional attribute, compared to most cereal grains, was their high-quality protein. Currently, oats are also known for their fiber content, specifically soluble dietary fiber. Traditionally, only the whole oat groat was used for human food, but recently grain fractions, such as the bran and even the hull, have entered the human food supply. The whole groat, or oatmeal, and the oat bran contain soluble dietary fiber; the hull contains negligible amounts.

The whole berry, or dehulled oat seed, is also known as the groat. The hull, which is about 25–30% of the grain, is removed by an impact machine to yield the oat berry after the grain has been cleaned and heated (Deane and Commers, 1986). Oat hulls, after cleaning, require special treatment to render them edible; this treatment includes fine grinding and possibly bleaching. There is some small use in baked goods. Rolled oats are normally prepared by flattening precut pieces of the groat, which produces a variety of somewhat differently cut and flattened products (McKechnie, 1983; Webster, 1986b). Oat bran is obtained by fractionation, usually by sieving, of coarse and fine particles from milled oat groat; it is usually 40–50% of the original whole groat. Oat milling and its history have been described in detail (Deane and Commers, 1986; Weaver, 1988). The primary aim of this chapter is to review and compare the macronutrient and selected micronutrient compositions of these oat fractions. The composition of oats also is compared to that of selected other grains. Because of the interest in oats as a source of dietary fiber, emphasis is given to the dietary fiber content and composition of oats and oat fractions. Since the dietary fiber content of a food may vary with the analytical method, this section is preceded by a brief discussion of the most common methods of dietary fiber analysis. This chapter is not intended to be an exhaustive review of all published data, and it complements other works (Matz, 1969; Youngs et al, 1982; Webster, 1986a; Peterson, 1992) on oats.

Proximate and Digestible Carbohydrate Analysis

MOISTURE

Environmental conditions at the time of harvest determine the initial moisture content of oats (McKechnie, 1983). Moisture is rarely >12% when harvest occurs in the warm summer months, as in the United States; it can be 20–25%, however, in Europe, where harvesting is done later in the fall. The moisture content of oat groats and bran is usually 7–9%; in the hulls it is lower, typically 4–8%.

PROTEIN

Recent proximate analyses, primarily from Scandinavia and Canada, indicate that the crude protein content of oat groats ranges from 11 to 24% (dry weight). Two cultivars, Hinoat and Goodland, both grown in Ottawa, ON, and developed for higher protein concentration, contained substantially more protein than the other varieties (Table 1).

Oat brans from six of the oat cultivars reported in Table 1 (Froker, Goodland, Hinoat, Selma coarse, Selma fine, and Sentinel) were also analyzed. The crude protein contents are higher in four of the brans than in the corresponding groats, lower in one (Hinoat), and not different in the Sang cultivar (Table 1). Bran from Canadian wild oats (*Avena fatua* L.) contains more protein than most of the other brans. Comparison of coarse and fine bran from the same milling of Selma (Frølich and Nyman, 1988) indicates that the finer product may be slightly lower in protein. The crude protein content reported for Mother's Oat Bran, a hot cereal commercially available in the early to mid 1980s, is at considerable variance with the manufacturer's label data and is probably in error. Alternatively, this bran contained other ingredients.

The crude protein content of oat hulls is low and is decreased further by the bleaching process used to whiten the product. Generally, the crude protein content of the oat groat is greater than that of whole barley (with hull) or rye meal and somewhat similar to that of whole wheat meal (Table 2). However, given the limited data in Table 2, it is not possible to distinguish any clear differences in the crude protein contents of oat, barley, rye, or wheat brans.

TABLE 1
Proximate Compositions
(% dry wt) of Different Fractions of Oat Grain

	Protein	Fat	Ash	CHO[a]	Data from
Groat					
Froker	16.2	8.4	2.4	NR[b]	Youngs, 1974
Goodland	21.5	8.8	2.7	NR	Youngs, 1974
Hinoat	24.4	6.6	2.3	66.7	Ma, 1983
Oxford	14.0	4.4	2.3	NR	Zarkadas et al, 1982
Sang					
1980	11.8	NA[c]	2.8	NR	Salomonsson et al, 1984
1981	12.4	NA	2.7	NR	Salomonsson et al, 1984
Selma	14.9	7.6	1.7	75.8	Frølich and Nyman, 1988
Sentinel	16.8	6.9	2.4	73.9	Ma, 1983
	12.4	6.6	2.2	NR	Zarkadas et al, 1982
SV78559	10.7	7.4	2.9	79.0	Aman and Hesselman, 1984
NS[d]	13.2	7.4	1.9	77.5	Sanchez- Marroquin et al, 1986
Bran					
Froker	20.0	9.6	4.5	NR	Youngs, 1974
Goodland	26.8	11.2	4.4	NR	Youngs, 1974
Hinoat	19.5	1.8	9.1	69.6	Ma, 1983
Sang, 1980	12.1	NA	8.1	NR	Salomonsson et al, 1984
Selma	17.1	7.7	2.5	72.6	Frølich and Nyman, 1988
Coarse					
Fine	16.8	8.6	2.3	72.3	Frølich and Nyman, 1988
Sentinel	17.5	2.7	7.5	72.3	Ma, 1983
Mother's	5.5	1.0	2.8	90.7	Chen et al, 1988
Quaker	19.9	7.6	3.1	69.4	Shinnick et al, 1988
Wild	25.9	12.2	4.4	57.5	Sosulski and Wu, 1988
Hull					
Unprocessed	3.9	1.3	4.7	90.1	Dougherty et al, 1988.
Selma	6.6	2.8	5.7	84.9	Frølich and Nyman, 1988
Bleached	0.8	0.1	2.3	96.8	Dougherty et al, 1988
	<0.5	~0.1	NA	NR	Lopez-Guisa et al, 1988
Processed[e]	2.8	1.8	NA	NR	Lopez-Guisa et al, 1988
NS	6.3	NA	6.5	NR	Garleb et al, 1988

[a]Carbohydrate.
[b]Not reported.
[c]Not analyzed.
[d]Not specified.
[e]Coated with 10% (by wt) starch.

LIPID

Except for one cultivar (Oxford), there is little variation in the crude fat content of the oat groat, which ranges from 6.6 to 8.8% (Table 1). Of the six brans produced from the corresponding groat samples for which data are reported, the brans from the Hinoat and Sentinel cultivars contain much less lipid than the groat, and the Froker and Goodland brans contain more. The Selma coarse bran has the same amount of lipid as the groat from which it originated, whereas the fine bran contains slightly more. Oat hulls contain little lipid, which is essentially extracted by bleaching.

The composition of the lipid in oats was reviewed by Youngs (1978). About 40% of the total lipid in both the groat and bran is triglyceride; free fatty acids and mono- and diglycerides account for 6–8%, sterols 2%, and phospholipids, 10–12%. About one quarter of the total lipid was not identified. Three fatty acids account for most of the neutral lipid (percent of total fatty acids): palmitic acid (16–24%), oleic acid (27–48%), and linoleic acid (33–46%).

Oat groats have more lipid than the other grains (Tables 1 and 2).

TABLE 2
Proximate Compositions (% dry wt) of Grains and Brans Other than Oats

	Protein	Fat	Ash	CHO[a]	Data from
Barley meal					
Alva, with hull	11.8	NA[b]	2.3	NR[c]	Salomonsson et al, 1984
Vanja, with hull	10.9	3.6	2.3	86.8	Aman and Hesselman, 1984
Betzes, with hull	10.5	3.0	2.7	83.8	Sumner et al, 1985
Scout, hull-less	19.9	2.0	2.2	75.9	Sumner et al, 1985
Barley bran					
Alva	11.7	NA	3.3	NR	Salomonsson et al, 1984
Rye meal					
Petkus II, 1980	12.9	NA	1.6	NR	Salomonsson et al, 1984
Petkus II, 1982	10.5	2.9	1.4	85.2	Aman and Hesselman, 1984
Rye bran					
Petkus II, 1980	15.1	NA	2.7	NR	Salomonsson et al, 1984
Wheat, whole meal					
SV 76427, winter	12.2	3.1	1.5	83.2	Aman and Hesselman, 1984
SV 75475, spring	13.2	3.1	1.4	82.3	Aman and Hesselman, 1984
Holme, winter	12.4	NA	1.7	NR	Salomonsson et al, 1984
Drabant, spring	13.0	NA	1.9	NR	Salomonsson et al, 1984
Wheat bran					
Holme, winter	12.8	NA	4.0	NR	Salomonsson et al, 1984
Drabant, spring	13.4	NA	3.0	NR	Salomonsson et al, 1984
AACC soft white	16.2	0.4	6.0	77.4	Chen et al, 1988

[a]Carbohydrate.
[b]Not analyzed.
[c]Not reported.

ASH

The ash content of oat groats ranges from 1.7 to 2.9% and is similar to the ash content of four of the bran samples (2.3–3.1%, Table 1). Ash was more concentrated in the six brans derived from the groats reported in Table 1. Some of these compositional differences may be a result of the use of different milling conditions to obtain the oat bran. Although bleaching decreased the ash content of the oat hulls, the ash concentrations of all samples in Table 1 are within the ranges of ash content for the oat groat and bran.

The average ash content of oat groats is similar to that of whole-seed barley meal but more than that reported for rye or whole-wheat meal (Tables 1 and 2). The limited data in Table 2 indicate that wheat bran, but not rye and barley bran, has more ash than oat bran.

CARBOHYDRATE BY DIFFERENCE

Carbohydrate was obtained by difference in a proximate analysis and thus would include low-molecular-weight carbohydrates, starch, and dietary fiber. It is about two thirds to three fourths (by weight) of the groat (Table 1). Except for the one commercial product, Mother's Oat Bran, carbohydrate in the oat bran is generally similar to or slightly lower than that in the oat groat. The hull is mainly dietary fiber, and this is reflected in its high content of carbohydrate. The cumulative effect of small differences in the crude protein, fat, and ash contents of oats compared to those in other grains makes the carbohydrate-by-difference generally higher in the other grains (Tables 1 and 2).

ANALYZED CARBOHYDRATE

Using direct methods of measurement rather than measurement by difference, Wood et al (1991) recently reported that the starch content of the oat groat ranged from 54.9 to 63.6% among 11 cultivars (Table 3). Analysis of the whole-seed meal (including hull) of the cultivar SV78559 gave 53.6% starch (Aman and Hesselman, 1984). However, the starch contents of whole-seed meals of the Sang cultivar over two harvests were lower (44.7 and 43.7%) (Salomonsson et al, 1984) than the starch content of the cultivars analyzed by either Wood et al (1991) or Aman and Hesselman (1984). The flour remaining after separating the bran was higher in starch (67.0–73.5%) than was the whole seed or groat (MacArthur and D'Appolonia, 1979b). Oat bran derived from the cultivars analyzed by Wood et al (1991) always contained less starch (41–52%) than the whole groat but more than the 35–38% measured by others (Salomonsson et al, 1984; Shinnick et al, 1988). Oat hulls contain little if any starch. Lopez-Guisa et al

(1988) measured 6.2% starch in cleaned but unbleached oat hulls, which was probably from contaminating endosperm, and were unable to detect any starch in the bleached product.

Unhulled barley and whole-meal rye contain about the same amount of starch as the whole oat groat (Tables 3 and 4). The concentration of starch in whole-meal wheat is usually higher. Dehulling and pearling of barley increases the concentration of starch (Table 4). Barley, rye, and wheat brans contain less starch than the unfractionated grains, but these starch concentrations are within the range for the oat brans analyzed by Wood et al (1991) (Tables 3 and 4). The certified soft white wheat bran available from the American Association of Cereal Chemists contains 20.6% starch and the certified hard red wheat bran 16.2%, indicating that wheat bran used in the United States may contain less starch than the Swedish wheats analyzed by Salomonsson et al (1984).

MacArthur-Grant (1986) reported that the sugar content of whole oats or groats was 0.9–1.4%, which is higher than the 0.5% determined for the Swedish oat cultivar Sang (Salomonsson et al, 1984) but comparable to the 1.1% found by Henry (1985). Low-molecular-weight carbohydrates are enriched in the bran compared to the groat (MacArthur-Grant, 1986). Shinnick et al (1988) measured 1.2% simple sugars, most of which (1.1%) was sucrose, in Quaker's Oat Bran. These are lower concentrations than those (2.4–3.4% total sugars, 1.7–2.7% sucrose) reported by MacArthur and D'Appolonia (1979a) for brans from three oat cultivars. Some of the difference in the total may be from the inclusion of more sugars by MacArthur and D'Appolonia in their analyses.

Lopez-Guisa et al (1988) measured 2.3% simple sugars in cleaned but unbleached oat hulls; none were detected in the bleached

TABLE 3
Starch Content (% dry wt) of Oat Groat and Bran[a]

| Cultivar | Starch Content | |
	Whole Groat	Bran
Marion	55.8	48.7
Capital	60.8	52.6
Woodstock	61.7	51.9
Sentinel	57.7	46.7
Ogle	60.4	50.9
Hinoat	54.9	41.1
Tibor	59.2	50.0
NO-1	55.1	47.1
Donald	63.6	52.5
OA 516-2	60.3	50.7
Harmon	59.8	49.5

[a]Data from Wood et al (1991).

product. These simple sugars may reflect contamination of the oat hulls with endosperm.

MacArthur-Grant (1986) reported the total free sugar and oligosaccharide concentrations in barley, rye, and wheat to be 2.6, 7.1, and 2.8%, respectively, twofold or more higher than measured in oat groat. The concentrations of low-molecular-weight carbohydrates in whole barley and wheat are similar to the concentrations reported to be in oat bran (MacArthur-Grant, 1986). Salomonsson et al (1984) detected more low-molecular-weight carbohydrates in whole-meal rye, barley, and wheat (1.0–1.2%) than in whole oat seed; Henry (1985) found a similar amount (1.4–2.7%).

TABLE 4
Starch Content (% dry wt) of Grains and Brans Other than Oats

	Whole-Seed Meal[a]	Bran	Data from
Barley			
Alva, 1980, with hull	58.8	45.9	Salomonsson et al, 1984
Alva, 1981, with hull	56.3	ND[b]	Salomonsson et al, 1984
Vanja, 1982, with hull	66.4	ND	Aman and Hesselman, 1984
Betzes, nonwaxy			
With hull	68.0	ND	Sumner et al, 1985
Dehulled	72.4	ND	Sumner et al, 1985
Pearled			
Medium extraction	74.8	ND	Sumner et al, 1985
Low extraction	78.7	ND	Sumner et al, 1985
Betzes isoline, waxy			
With hull	56.7	ND	Sumner et al, 1985
Dehulled	63.0	ND	Sumner et al, 1985
Pearled			
Medium extraction	67.2	ND	Sumner et al, 1985
Low extraction	72.4	ND	Sumner et al, 1985
Scout, nonwaxy, hull-less	60.4	ND	Sumner et al, 1985
Pearled			
Medium extraction	67.4	ND	Sumner et al, 1985
Low extraction	70.6	ND	Sumner et al, 1985
Rye			
Petkus II, 1980	57.7	45.3	Salomonsson et al, 1984
Petkus II, 1982	63.2	ND	Aman and Hesselman, 1984
Wheat			
Holme, 1980	62.8	38.3	Salomonsson et al, 1984
Holme, 1981	62.1	ND	Salomonsson et al, 1984
SV 76477, 1980	63.7	48.1	Salomonsson et al, 1984
SV 76477, 1981	58.4	ND	Salomonsson et al, 1984
Drabant, 1980	60.9	50.2	Salomonsson et al, 1984
Drabant, 1981	60.6	ND	Salomonsson et al, 1984
SV 75475, 1982	68.8	ND	Aman and Hesselman, 1984
SV 76427, 1982	69.0	ND	Aman and Hesselman, 1984

[a] Some barley samples were dehulled or pearled, as indicated.
[b] No data.

Minerals

The mineral content of oats varies significantly among different cultivars (Table 5) and is probably affected by differences in the cultivar, soil mineral content, fertilizer usage, and other growing conditions. In all types of oat samples, the major minerals are calcium, magnesium, phosphorus, and potassium. Data from the study by Yli-Halla and Palko (1987) suggest that the soil in Finland influenced only the calcium and sodium contents. Oat hulls are sufficiently high in calcium (Frølich and Nyman, 1988) to be potentially a nutritionally significant source, provided the calcium is bioavailable. However, most of the ash appears to be lost with bleaching of hulls (Table 1). The whole groat provides nutritionally significant amounts of calcium, iron, and zinc and is usually low in sodium (Table 5). The mineral content of the coarse oat bran was less than that of the oat groat, and that of the fine bran was still lower (Frølich and Nyman, 1988). This contrasts with other data (Table 1), in which the average ash in the groat (2.5%) was less than half that in the bran (5.8%) (omitting data of Frølich and Nyman, 1988).

The mineral concentration in oats is comparable to that of a large series of composites of hard red winter wheat (Pomeranz and Dikeman, 1983). The average calcium content of the hard wheat samples was 487 mg/g; iron was 50 mg/g and zinc, 28 mg/g. Some cultivars of oats, therefore, may provide a larger amount of nutritionally important minerals than hard red wheat. Phytic acid, which has been associated with reduced mineral bioavailability, is present in oats at a

TABLE 5
Mineral Concentration (μg/g of dry wt) of Oat Grain and Fractions

| | | Whole Grain[b] | | | | | |
| | | 35 Samples | | | Selma Oat Fractions[c] | | |
Mineral	171 Groat Samples[a]	Grown in Acid Sulfate Soils	19 Samples Grown in Other Soil	Hull	Groat	Coarse Bran	Fine Bran
Calcium	70–1,800	410–820	500–1,600	960	620	267	186
Copper	3–8	4–15	4–15	2	5	1	1
Iron	NA[d]	57–300	74–310	26	45	17	13
Magnesium	1,000–1,800	1,200–1,700	1,200–1,600	490	1,400	658	406
Manganese	22–79	48–200	39–190	37	51	18	13
Phosphorus	2,900–5,900	2,900–4,500	3,300–4,400	390	3,400	1,496	952
Potassium	3,100–6,500	3,800–6,200	4,500–5,200	7,700	4,200	1,890	1,134
Sodium	40–600	14–770	15–240	90	15	5	4
Zinc	21–70	28–54	29–50	12	35	12	8

[a]Data from Morgan (1968).
[b]Data from Yli-Halla and Palko (1987).
[c]Oat grain was 30% husk and 70% kernel (data from Frølich and Nyman, 1988).
[d]Not analyzed.

level (0.7%) comparable to that in wheat and is enriched in the bran (1.0–1.2%) (Frølich and Nyman, 1988).

Amino Acid Composition

Several analyses of the amino acid composition of oat groats, bran, and hull have been reported (Tables 6 and 7). Pomeranz et al (1973) compared the amino acid composition of commercially milled oats with those of oats obtained by hand separation of the Orbit cultivar and of 11 species from the U.S. Department of Agriculture (USDA) World Collection. The commercial sample was a blend of oats from the North Central United States that were obtained from Quaker Oats Co. The cultivars from the World Collection are not grown commercially in the United States because of their poor agronomic characteristics but are potentially useful as high-protein breeding stock. Cultivars specified are: Condor, from Scotland (Draper, 1973); Coker 227, harvested in 1976 and 1977 in the state of Georgia (Morey and Evans, 1983); and Hinoat, harvested in 1978, and Sentinel, harvested in 1980, in Ottawa, Ontario, Canada (Ma, 1983). The other Sentinel and the Oxford cultivar were grown in Canada, but location and year were not specified (Zarkadas et al, 1982). Thomke (1988) reported the amino acid composition of a variety of oat groats and cereals, most of which were originally reported in the Norwegian literature.

The amino acid compositions of the groat samples obtained by hand separation (Orbit and USDA World Collection data) are generally similar to the amino acid composition of the groats obtained by milling (Table 6). The range of amino acid concentrations from analyses of the individual cultivars is in general agreement with the range reported by Thomke, and the mean amino acid composition of the 11 species from the World Collection is within the range of groat amino acid data reported in Table 6.

Nutritionists in the United States seeking initial information about the nutrient composition of a foodstuff may begin with one of the USDA tables. Many of those data are old, and cultivar breeding over 25–50 years might have made a significant impact on nutrient composition. In many instances, the range or the mean of amino acid contents reported in USDA Handbook no. 4 (Orr and Watt, 1968) is lower than either the range of the individual cultivar analyses in Table 6 or the range from the Thomke (1988) review; both the range and average of the USDA data are lower than the others for arginine, aspartic acid, serine, glutamic acid, cystine, and leucine. The upper ranges in the USDA table for alanine, valine, and isoleucine are not found among the other data in Table 6. Nearly all of the more recent

TABLE 6
Amino Acid Composition (g/100 g of protein) of the Oat Groat[a]

	Mixed	Condor	Coker 227 1976	Coker 227 1977	Hinoat	Orbit	Oxford	Sentinel		USDA Mean[b]	Thomke Review Range	USDA Data Mean	USDA Data Range
	A	B	C	C	D	A	E	E	D	A	F	G	G
Lys	3.9	4.2	4.0	4.1	3.4	4.5	4.6	3.9	4.1	3.8	3.4–4.7	NA[c]	NA
His	2.3	1.7	1.7	1.7	2.1	2.4	2.6	2.7	2.2	2.2	2.0–2.8	1.7	1.2–2.4
Arg	6.2	6.8	6.9	7.0	6.0	6.8	6.1	6.2	5.9	6.7	6.0–9.1	6.1	4.1–7.5
Asp	9.0	8.5	7.9	8.1	7.3	8.7	9.3	9.4	7.8	8.3	7.8–9.5	3.9	3.0–4.5
Thr	3.1	3.3	3.5	3.7	2.8	3.4	3.6	3.9	3.1	3.3	3.2–3.8	3.1	2.1–3.6
Ser	3.9	5.2	4.9	5.0	4.1	4.6	3.7	4.9	4.2	4.2	4.4–6.2	3.7	2.4–4.9
Glu	22.4	21.0	21.1	21.8	18.6	21.7	22.1	20.6	19.5	22.6	18.4–24.3	18.8	17.1–20.5
Gly	5.0	5.3	4.6	4.7	4.0	5.2	5.3	5.4	4.6	4.9	4.7–5.5	4.2	3.1–5.2
Ala	5.0	5.1	4.6	4.7	4.0	5.0	4.9	5.1	4.6	4.7	4.6–5.5	5.7	4.1–7.0
Cys/2	2.0	1.6	1.8	1.9	1.7	2.1	1.7	1.8	2.3	2.3	1.7–1.8	2.0	1.3–3.0
Val	5.7	4.6	5.3	5.4	4.7	5.5	5.8	5.8	4.9	5.5	4.6–6.8	5.6	3.8–7.0
Met	2.5	1.4	1.8	1.8	1.1	2.2	3.8	3.7	1.1	2.9	1.5–1.9	1.4	1.0–2.3
Ile	4.3	3.5	4.0	4.1	3.5	3.9	4.3	4.3	3.8	3.9	3.6–4.4	4.8	4.1–5.9
Leu	7.4	6.8	7.5	7.7	6.9	7.6	7.9	8.5	7.4	7.3	6.9–8.9	3.4	1.9–5.1
Tyr	2.5	3.9	3.7	3.8	3.1	3.0	2.0	3.0	3.1	3.3	3.0–4.6	3.4	1.5–4.5
Phe	5.5	5.0	5.5	5.6	5.7	5.2	5.1	5.1	5.9	5.3	4.6–6.3	5.0	4.2–7.2
Pro	6.2	6.7	5.1	5.4	4.4	5.5	6.7	5.9	5.0	6.1	4.4–5.9	5.3	4.1–6.5

[a]Data from: A, Pomeranz et al (1973); B, Draper (1973); C, Morey and Evans (1983); D, Ma (1983); E, Zarkadas et al (1982); F, Thomke (1988); G, Orr and Watt (1968).
[b]USDA World Collection.
[c]Not analyzed.

data were collected by an automatic amino acid analyzer after hydrolysis with hydrochloric acid; some early data on amino acid content were obtained by microbiological assays. Thus, it is possible that some of the discrepancies between recent and earlier data are the result of methodological differences.

Several of the data for the amino acid compositions of the oat hull and the oat bran (Table 7) may not be representative of commercially available products. The Condor and USDA World Collection oat hulls were obtained by manual dissection (Draper, 1973), as was the bran from the Orbit cultivar (Pomeranz et al, 1973). To obtain the Hinoat and Sentinel oat bran, aliquots of ground groats were defatted by Soxhlet extraction and then extracted with different concentrations of base (NaOH, 0.005–0.05N); the bran was recovered as the residue following centrifugation or filtration through cheesecloth (Ma, 1983).

The amino acid contents of the two alkali-insoluble brans, Hinoat and Sentinel (Ma, 1983), are comparable to or higher than those of the corresponding groats, except for three amino acids; both brans contain less cystine and glutamic acid and the Sentinel bran contains less phenylalanine than the groats from which they originated (Table 7). However, the amino acid composition of the bran obtained by hand dissection of the Orbit cultivar does not show much difference from its respective groat; there are slightly higher amounts of serine,

TABLE 7
Amino Acid Composition (g/100 g of protein) of the Oat Bran and Hull

| | Oat Bran | | | Oat Hull | | |
	Hinoat A[a]	Orbit B	Sentinel A	Mixed B	Condor C	USDA World Collection Mean B
Lys	5.3	4.1	5.4	4.9	3.9	5.6
His	2.7	2.2	2.6	2.4	NA[b]	2.1
Arg	8.2	6.8	7.8	6.8	4.5	4.1
Asp	8.8	8.6	8.7	10.5	8.4	15.6
Thr	4.0	3.4	3.6	4.1	3.9	4.8
Ser	5.1	4.8	4.2	4.6	5.4	4.6
Glu	19.1	21.1	17.2	20.3	8.7	14.5
Gly	5.5	5.4	5.6	6.1	5.6	5.8
Ala	5.9	5.1	5.7	5.4	5.5	7.0
Cys/2	1.5	2.4	1.6	0.5	NA	0.2
Val	6.0	5.5	6.1	6.4	5.1	6.4
Met	1.6	2.1	1.1	1.5	1.4	2.3
Ile	3.9	3.8	3.8	4.5	2.9	3.9
Leu	8.0	7.4	7.4	7.8	6.1	7.2
Tyr	3.3	3.5	3.3	2.9	2.6	2.3
Phe	6.0	5.1	5.6	5.3	6.3	4.3
Pro	5.0	6.2	4.9	2.4	3.6	4.8

[a]Data from: A, Ma (1983); B, Pomeranz et al (1973); C, Draper (1973).
[b]Not analyzed.

glycine, alanine, cystine, tyrosine, and proline in the bran, compared to the groat. These data indicate that alkali-insoluble bran may be higher in amino acid concentration than bran obtained by hand separation.

The concentrations of several amino acids are lower in the hull of the Condor cultivar than the mean amino acid concentrations of the 11 species from the USDA World Collection, even though all samples were obtained by hand separation. The Condor oat hull contains less lysine, aspartic acid, threonine, glutamic acid, glycine, alanine, valine, methionine, isoleucine, leucine, and proline than the average amounts of the amino acids in the World Collection. The commercially milled hull contains less lysine, aspartic acid, threonine, alanine, methionine, and proline than the World Collection oat hulls.

All of the data in Tables 6 and 7 are expressed on the basis of protein content. The protein concentration of oat bran is similar to that of oat groat (Table 1), indicating that many oat brans and groats contain similar amounts of amino acids. In contrast, the protein content of oat hulls is low, making them poor sources of amino acids. Bleaching removes virtually all of the protein from hulls (Table 1).

Traditionally, the quantity and quality of the oat groat protein,

TABLE 8
Amino Acid Composition (g/100 g of protein) of Other Grains and Soybean

	Barley[b]	Rice[c]	Rye[d]	Soybean	Triticale[e]	Wheat[f]
	A	B	A	A (n = 1)	C	A
Lysine	3.8	3.7	3.5	7.0	3.6	2.8
Histidine	1.7	2.3	1.7	2.3	2.3	1.7
Arginine	5.3	8.3	5.3	7.8	5.1	4.9
Aspartic acid	6.3	8.6	6.6	11.4	6.2	4.9
Threonine	3.5	3.5	3.2	4.3	3.2	2.9
Serine	4.3	5.0	4.5	5.5	4.6	4.7
Glutamic acid	22.7	16.7	25.3	19.5	26.3	31.0
Glycine	4.1	4.5	3.9	4.5	4.2	3.9
Alanine	4.0	5.4	3.8	4.6	4.0	3.4
Half-cystine	1.2	2.3	1.3	1.2	2.5	1.5
Valine	4.7	5.9	4.4	5.0	4.6	4.2
Methionine	1.7	2.3	1.8	1.8	1.6	1.8
Isoleucine	3.5	4.1	3.4	5.0	3.7	3.6
Leucine	6.9	7.9	6.1	8.0	6.3	6.7
Tyrosine	3.3	5.0	2.7	4.1	3.0	3.3
Phenylalanine	5.0	5.0	4.7	5.5	4.5	4.7
Proline	10.1	4.4	10.4	5.2	9.4	10.1

[a]Data from: A, Morey and Evans (1983); B, (Mosse et al, 1988b); C, Mosse et al (1988a).
[b]Mean of one cultivar, Volbar, grown in 1976 and 1977.
[c]Mean of eight different varieties.
[d]Mean of three different cultivars.
[e]Mean of 19 samples of seven varieties.
[f]Mean of six samples of five cultivars.

and not the dietary fiber, was the nutritional attribute of oats that was stressed (McKechnie, 1983). Indeed, compared to the amino acid composition of rye, triticale, and wheat, the oat groat has greater amounts of several amino acids (Tables 6–8). Levels of lysine, arginine, aspartic acid, glycine, alanine, cystine, valine, leucine, tyrosine, and phenylalanine in rye, triticale, and wheat are either lower than, or at the low end of the range of, the concentrations of these amino acids in oat groats. The dominant amino acids in these three grains are glutamic acid and proline. The concentrations of several amino acids in soybeans are either similar to or greater than those in oat groats. Compared to oat groats, rice has more arginine, alanine, valine, methionine, and tyrosine and less glutamic acid and cystine. The distribution and concentrations of amino acids in barley are somewhat similar to those in oat groats; barley has less arginine, aspartic acid, and cystine and more glutamic acid and proline (Tables 6–8). The rye, barley, soybean, and wheat varieties summarized in Table 8 were grown in the United States in Georgia in 1976 and 1977; the wheat varieties were all soft red winter cultivars (Morey and Evans, 1983). The rice varieties were from Montpellier, France (Mosse et al, 1988b). The triticale was also grown in France (Mosse et al, 1988a).

Dietary Fiber

METHODS OF ANALYSIS

The measured dietary fiber content of oats and oat products may vary with the method of fiber analysis, which limits the comparison and interpretation of data. Methods developed for dietary fiber analysis have proliferated recently (Marlett, 1990a). Because of the complexity of the plant matrix and the heterogeneity of the components of dietary fiber, a method developed for one type of sample may not accurately analyze fiber in another. Frequently, these limitations are discovered only after a method is published. Most methods have continued to evolve, incorporating changes that improve analytical reproducibility or speed the procedure. The net result is that fiber values should be interpreted with careful reference to the method, and the modification of the method, that was used (Marlett, 1990a).

Methods for measuring fiber in human foods have been reviewed recently (Pilch, 1987; Marlett, 1990a,b). Briefly, methods can be classified on the basis of analytical approach as either gravimetric or chemical. Gravimetric methods that have been used to measure dietary fiber in grains include those developed by Van Soest and Wine (1963) (the neutral detergent fiber [NDF] method), Asp et al (1983),

Mongeau and Brassard (1986, 1989), and Prosky et al (1985, 1988, 1992) (the AOAC method). Chemical methods of fiber analysis developed by Theander (Theander and Aman, 1979; Theander and Westerlund, 1986) and Englyst (Englyst et al, 1982; Cummings and Englyst, 1987) have been used to analyze grains. Unless precautions are taken to recover polysaccharides solubilized by the detergent solution, the NDF method underestimates total fiber (Mongeau and Brassard, 1986). However, the NDF method may overestimate total fiber if starch is not adequately removed by the amylase treatment (Robertson and Van Soest, 1981).

In gravimetric methods, nonfiber constituents of food are partially removed, the fiber polysaccharides are precipitated with ethanol, and the residue is dried and weighed. Additional analyses on the residue then correct for incompletely removed nonfiber components. Most commonly, as in the AOAC procedure, protein and ash are the corrections applied to the residue (Prosky et al, 1988). Nitrogen is converted to crude protein with the value of 6.25, and this may inflate protein correction, particularly if the residue is high in protein.

In chemical methods of analysis also, nonfiber constituents of foods are partially removed. However, aliquots of the residue are then analyzed, after acid hydrolysis or solubilization, for the monomeric constituents of the fiber polysaccharides (Marlett, 1990a). Uronic acids, the major sugars in pectins but usually arising in cereals from acidic pentosan, are measured either colorimetrically or by quantitating the carbon dioxide released by decarboxylation. Neutral sugars are measured indirectly by colorimetric methods or directly by high performance liquid chromatography (HPLC) or by gas chromatography. Lignin is recovered as Klason lignin, that is, material insoluble after treatment with 72% sulfuric acid (Theander and Aman, 1981).

One chemical method, which is a modification of the pioneering method of Southgate (1969, 1981), does not measure lignin (Cummings and Englyst, 1987). Grains are the major source of lignin in the human food supply. Inclusion of lignin as dietary fiber is controversial because there are no accurate methods for its measurement, although available data suggest that Klason lignin in grains does reflect true lignin (Marlett, 1990a). However, several components in foodstuffs, such as tannins and saponins, are recovered in the Klason lignin (Robertson and Van Soest, 1981). We have found that protein also may contaminate the Klason lignin, particularly if the residue is not fully neutralized before it is collected (Marlett, 1990b). Most investigators agree that lignin should be included as part of dietary fiber (Robertson and Van Soest, 1981; Theander and Aman, 1981; Bingham, 1987; Marlett, 1990a). Phenolics, components of lignin, exert a variety of physiological effects, which may include carcinogenic and anticarcinogenic activities (Newmark, 1987); they may be responsible for some of the effects currently attributed to fiber (Marlett, 1990a).

Soluble and insoluble fiber fractions are usually measured by chemical methods (Marlett, 1990a), but two gravimetric methods also have been proposed. Mongeau and Brassard (1986) analyzed separate aliquots of a sample for soluble and insoluble fiber, the latter obtained by the NDF method. Our data indicate that the NDF method does not always recover an insoluble fraction of many foods that is comparable to that measured by the Theander chemical method (Marlett, 1990a). With a more vigorous extraction of the soluble fraction than used in most fiber methods (for a review, see Marlett, 1990a), Mongeau and Brassard (1989) recovered total dietary fiber, as the sum of this soluble fraction and NDF. Their total dietary fiber was similar to the total dietary fiber analyzed by the AOAC procedure (Prosky et al, 1985).

The AOAC method has undergone three collaborative trials to evaluate a modification designed to fractionate fiber (Prosky et al, 1988, 1992, 1993); the results of the first two studies indicated that the soluble fiber fraction could not be measured reproducibly by this gravimetric approach. The recommendation from the third study was that the method for the determination of soluble dietary fiber be given official first action adoption by the AOAC International.

All gravimetric procedures (except NDF) and some chemical procedures recover fiber by precipitating the polysaccharides with ≈80% ethanol. It is known that one type of hemicellulose does not readily precipitate in ethanol (Larm et al, 1975). It is also possible that other polysaccharides may not completely flocculate in the ethanol step; these polymers might escape measurement as fiber in either chemical or gravimetric procedures unless precipitation with other concentrations of ethanol is successful (Siddiqui and Wood, 1974). Gravimetric and some chemical methods do not include a step to extract simple sugars endogenous to the food or the mono- or disaccharides added during processing. When these sugars are present in very high concentrations, they may coprecipitate with the fiber polysaccharides when the sample is treated with ethanol and be measured as dietary fiber (Marlett and Navis, 1988).

Incomplete acid hydrolysis may be a source of underestimating fiber in chemical methods (Marlett, 1990a). Complete solubilization of polysaccharides certainly does not mean that they have been completely hydrolyzed (Marlett, 1990b). We detect cellobiose, which is possible by HPLC but not with the gas chromatographic methods used in dietary fiber analysis, as an index of incomplete hydrolysis and currently use a Saeman hydrolysis for both soluble and insoluble fiber fractions (Marlett, 1990b). The hydrolytic degradation losses that may also occur are estimated by determining hydrolysis losses of a standard sugar mixture.

Methodology-based differences in dietary fiber values can be verified only by analyzing the same sample by the different methods. Some differences for the same foodstuff (but different sample) may be a consequence of growing conditions or cultivars. Consistent trends in differences found with different analytical methods over a period of time by several laboratories would indicate, however, that the differences are probably based on differences in methodology.

(1→3),(1→4)-β-D-GLUCAN

β-Glucan is a cell wall polysaccharide that is present in oats and barley in much greater concentrations than in other grains (Table 9)

TABLE 9
β-Glucan Content (% dry wt) of Grains and Cereal Products

Sample	Number Analyzed[a]	Mean	Range	Data from
Oat[b]	2	5.7	4.8–6.6	Prentice et al, 1980
Oat[b]	2	3.4	NR[c]	Henry, 1985
Oat[b]	1	3.0	NR	Aman and Hesselman, 1985
Oat groat	11	5.1	3.9–6.8	Wood et al, 1991
Oat groat	97	NR	3.2–6.3	Welch and Lloyd, 1989
Oat breakfast cereals	3	3.8	2.5–4.6	Carr et al, 1990
Oat bran	11	7.4	5.8–8.9	Wood et al, 1991
Oat bran	1	7.3	NR	Carr et al, 1990
Oat bran breakfast cereals	3	4.3	1.9–8.9	Carr et al, 1990
Malting barleys[b]	19	6.4	4.5–8.2	Prentice et al, 1980
Feed barleys[b]	10	6.1	5.1–7.2	Prentice et al, 1980
Barley[b]	2	4.4	3.9–4.8	Henry, 1985
Barley[b]	2	3.8	3.7–3.8	Aman and Hesselman, 1985
Rice	2	0.1	NR	Henry, 1985
Rye[b]	3	2.4	1.9–2.9	Prentice et al, 1980
Rye[b]	2	1.9	1.7–2.0	Henry, 1985
Rye[b]	1	1.3	NR	Aman and Hesselman, 1985
Sorghum[b]	1	1.0	NR	Prentice et al, 1980
Triticale[b]	1	1.2	NR	Prentice et al, 1980
Triticale[b]	2	0.6	0.4–0.8	Henry, 1985
Triticale[b]	1	0.5	NR	Aman and Hesselman, 1985
Wheat[b]	2	1.4	NR	Prentice et al, 1980
Wheat[b]	2	0.7	0.6–0.7	Henry, 1985
Wheat[b]	1	0.5	NR	Aman and Hesselman, 1985

[a]Samples or cultivars.
[b]Methods indicated that samples were ground and analyzed. No milling or dehulling is described. It is assumed these are samples of the whole grain, including hull, if appropriate.
[c]Not reported.

and has been proposed as one of the factors in oats and barley responsible for their ability to lower blood cholesterol concentrations. Until recently, a separate determination of β-glucan was not included as part of dietary fiber analysis; instead, it was measured as part of the total fiber in gravimetric procedures and as part of the glucose in chemical methods. The β-glucan content of the whole oat kernel and groat ranges from 3.0 to 6.8%. In the bran, the content is higher, ranging from 5.8 to 8.9% in the 11 cultivars analyzed by Wood et al (1991). Breakfast cereals containing oats have a wide range of β-glucan concentrations.

The β-glucan concentration in barley is usually comparable to the range reported for oat groat. Descriptions for several of the samples reported in Table 9 did not clearly indicate whether hulls were removed; therefore, it has been assumed that the whole grain (including hull, if appropriate) was analyzed. The β-glucan concentration would be higher on the dehulled product. Rye contains 1–3% β-glucan, whereas rice, sorghum, triticale, and wheat samples had <1% β-glucan. Most methods of β-glucan analysis use enzymatic hydrolysis to degrade the polymer and then measure the glucose. However, some methods extract the β-glucan before analysis. Incomplete extraction would lead to an underestimation of β-glucan. Major differences in β-glucan content (and dietary fiber) among oats and barley have a genetic basis, but environment also may have an effect (Newman and Newman, 1991; Wood et al, 1991).

COMPOSITION OF SOLUBLE AND INSOLUBLE DIETARY FIBER IN OATS

In the Whole Oat Kernel and Oat Groat

The whole oat kernel contains more fiber than the groat because it includes the hull (Table 10). The total dietary fiber content of oat groats ranged from 7.1 to 12.1% in five reported analyses. Two values (Cummings and Englyst, 1987) are probably low because lignin is not included. The oat groats analyzed by Frølich and Nyman (1988) using the method of Asp et al (1983) had less fiber in the soluble and more in the insoluble fraction. The method of Asp et al (1983) includes three enzymatic steps to remove starch and protein, one of which is a pepsin incubation at pH 1.5. It is possible that the combination of these three treatments rendered more of the total fiber insoluble (Marlett et al, 1989). Alternatively, the differences are a consequence of different growing conditions or cultivars.

Of the total fiber in milled oats, 40–50% was recovered in the soluble fraction by three laboratories; 56–69% of the total was soluble when the Englyst method was used (Table 10). This difference is a

TABLE 10

Composition (% dry wt) of the Soluble and Insoluble Dietary Fiber in Different Fractions of Oat Grain

	Soluble				Insoluble					Total Fiber	Data from[a]
	Neutral Sugars	Uronic Acids	β-Glucan	Total	Neutral Sugars	Uronic Acids	β-Glucan	Klason Lignin	Total		
Whole kernel											
Sang	1.4	0.1[b]	INC[c]	NC[d]	20.5	1.1[e]	INC	5.3	NC	28.3	E
Groat											
Oatmeal	1.0	0.1	3.8	4.9	3.0	0.3	0.6	3.3	7.2	12.1	F
Selma	2.9	0.1	INC	3.0	5.6	0.4	INC	2.0	8.0	11.0	C
Coarse	4.7	0.1	INC	4.8	2.9	0.1	INC	NA[f]	3.0	7.8	B
Porridge	3.9	0.1	INC	4.0	3.0	0.1	INC	NA	3.1	7.1	B
Rolled oats	5.3	0.1	INC	5.4	3.9	0.2	INC	1.0	5.1	10.5	A
Bran											
Sang	1.1	0.1	INC	NC	21.5	1.3	INC	7.3	NC	31.2	E
NS[g]	1.6	0.1	5.3	7.0	5.0	0.3	2.2	3.8	11.3	18.3	F
Selma											
Coarse	5.6	0.1	INC	5.7	5.9	0.5	INC	3.0	9.4	15.1	C
Fine	5.2	0.1	INC	5.3	6.9	0.4	INC	3.0	10.3	15.6	C
NS	7.7	0.1	INC	7.8	6.3	0.4	INC	1.6	8.3	16.1	A
Bran plus germ	8.3	0.1	INC	8.4	5.3	0.2	INC	NA	13.7	22.1	B
Hull											
Unprocessed	0.3	0.1	NA	0.4	56.7	1.8	NA	20.0	78.5	78.9	C
Processed[h]	0.5	0.1	0	0.6	60.4	1.5	0	17.5	79.4	80.0	D
Bleached	3.7	0.2	0	3.9	69.1	1.1	0	11.4	81.6	85.5	D

[a]A, Anderson and Bridges (1988); B, Cummings and Englyst (1987); C, Frølich and Nyman (1988); D, Lopez-Guisa et al (1988); E, Salomonsson et al (1984); F, Shinnick et al (1988).
[b]Unavailable; included with insoluble fiber uronic acids.
[c]Not analyzed separately but included in neutral sugars.
[d]Not calculable, since soluble uronic acids are included in the insoluble uronic acids.
[e]Includes uronic acids from soluble fiber fraction.
[f]Not analyzed.
[g]Not specified.
[h]Coated with 10% (by weight) starch.

consequence of excluding lignin, since without lignin both the total and the insoluble fractions are smaller. The net effect is an increase in the percentage of total fiber that is apparently soluble. The absolute amounts of soluble fiber obtained by the Englyst method, however, are similar to those determined by other methods.

Only one of these studies (Shinnick et al, 1988) specifically measured β-glucan. Fourteen percent of the total β-glucan in the milled oat groat was measured in the insoluble fiber fraction when the Theander method (Theander and Westerlund, 1986) of analysis was used to obtain the soluble and insoluble fractions (Shinnick et al, 1988).

In Oat Bran and Hull

Oat bran contains 15–19% total dietary fiber; 34–48% of the total measured as soluble fiber (Table 10). Most of the bran samples contain about 50% more fiber than the milled oats. The oat bran with the highest dietary fiber also contained the highest Klason lignin (Shinnick et al, 1988), possibly because it was contaminated with protein (Marlett, 1990b), but the elevated lignin does not account for all the increased dietary fiber. The effect of adding an unknown amount of oat germ to oat bran (which normally contains some germ) was to increase both the total and the insoluble fiber (Table 10). The fiber of oat bran contains about the same proportion of β-glucan as the milled oats (41 and 36%, respectively), although a larger proportion (29%) of the total β-glucan in the bran is insoluble compared to that in the groat (14%).

The oat hull is a concentrated source of insoluble fiber (Table 10). When other constituents, e.g., protein, fat, and ash, were removed with bleaching, the concentration of total fiber increased. There is no mixed-linkage β-glucan in the oat hull.

None of the oat fractions contain significant amounts of uronic acids.

Effect of Method of Analysis on Dietary Fiber in Oat Bran

Marlett et al (1989) recently compared the distribution of total fiber between the soluble and insoluble fractions in oat bran obtained by a modification (Shinnick et al, 1988) of the Theander method A (Theander and Westerlund, 1986) with the distribution obtained by two additional modifications. In one modification, a protease digestion step similar to the one used in the AOAC total fiber method was added, and in the other a pepsin digestion step was added, similar to that used in the gravimetric method of Asp et al (1983). The additional treatments had no effect on the distribution of the fiber-derived neutral sugars in oat bran, but they increased the proportion of the total β-glucan that was extracted into the soluble fraction (Table 11). The addition of the pepsin digestion step increased the value for Klason lignin, possibly by contamination with precipitated protein.

Graham et al (1988) reported that neutral sugars recovered as soluble fiber decreased in several foods, including oat bran, if the sample was incubated at pH 1.5, 38°C, or treated with absolute ethanol, 96°C, without any enzyme steps to hydrolyze starch. The enzymatic treatments to hydrolyze starch also disrupt the plant cell matrix; thus, it is possible that some of the decrease in soluble fiber yields observed by Graham et al (1988) were due to the elimination of those steps.

Neutral Sugar Composition of Dietary Fiber Fractions of Oat Grain

The predominant neutral sugar in the soluble fiber of oat groat or bran is glucose because the major polymer in the soluble fraction is β-glucan (Table 12). The Sang cultivar samples grown in Sweden had less glucose and more arabinose and xylose than the others. Since these samples were obtained by an experimental milling process, it is possible that the hull was not completely removed, which would explain these differences.

The neutral sugars in the insoluble fraction of the groat or bran, from highest to lowest concentration, are glucose, xylose, and arabinose. The glucose is from β-glucan and cellulose and possibly small amounts of glucomannan. Since amounts of glucomannan are likely to be small, only cellulose should be a significant source of glucose in insoluble fiber. Theander and Aman (1981) assumed that all of the glucose in the insoluble fraction is from cellulose except when β-glucan is present. Xylose and arabinose generally arise from arabinoxylan, possibly

TABLE 11
Effect of Method of Dietary Fiber Analysis on Content (% of total fiber)
of Soluble and Insoluble Fiber in Oat Bran[a]

Method of Analysis	Neutral Sugars	β-Glucans	Uronic Acids	Klason Lignin	Total
Theander method A					
Soluble	1.0	5.4	0.1	0	6.5
Insoluble	3.8	2.2	0.4	3.5	9.9
Method A plus protease step					
Soluble	1.1	5.9	0.2	0	7.2
Insoluble	4.4	1.8	0.4	2.3	8.9
Method A plus pepsin step					
Soluble	1.2	6.1	0.2	0	7.5
Insoluble	4.1	1.3	0.4	5.2	11.0

[a]Modified from data of Marlett et al (1989).

TABLE 12
Neutral Sugar Composition of Soluble and Insoluble Dietary Fiber
from Different Fractions of Oat Grain

	Soluble, % of neutral sugars						Insoluble, % of neutral sugars						Data from[b]
	Glc[a]	Xyl	Ara	Man	Gal	Rha	Glc	Xyl	Ara	Man	Gal	Rha	
Groat													
Oatmeal	87	5	7	1	0	0	42	29	20	4	5	ND[c]	E
Sang	78	6	8	2	6	T[d]	45	45	7	1	2	T	D
Selma	88	3	3	3	3	NA[e]	54	25	16	3	2	NA	B
Oatmeal, coarse	94	2	2	T	2	T	34	38	28	T	T	T	A
Porridge oats	87	5	5	T	3	T	37	33	23	3	3	T	A
Bran													
Quaker	81	5	6	3	5	ND	45	29	18	3	5	ND	E
Sang	61	13	13	2	11	T	45	45	7	1	2	T	D
Selma, coarse	91	2	3	Tr[f]	4	NA	49	30	18	2	1	NA	B
Selma, fine	86	4	4	2	4	NA	54	23	17	3	3	NA	B
Bran plus germ	92	1	5	Tr	2	T	35	39	25	Tr	Tr	T	A
Hull													
Selma, unprocessed	67	Tr	Tr	Tr	33	NA	52	39	6	Tr	3	NA	B
Processed[g]	61	16	14	0	9	ND	45	49	4	0	2	ND	C
Bleached	8	77	14	0	1	ND	49	45	6	0	Tr	ND	C

[a]Glc = glucose, Xyl = xylose, Ara = arabinose, Man = mannose; Gal = galactose, Rha = rhamnose.
[b]A, Cummings and Englyst (1987); B, Frølich and Nyman (1988); C, Lopez-Guisa et al (1988); D, Salomonsson et al (1984); E, Shinnick et al (1988).
[c]Not detected. Rha and Gal coelute by high-performance liquid chromatographic method used.
[d]Trace, concentration not specified.
[e]Not analyzed.
[f]Trace, <0.05 g/100g.
[g]Coated with 10% (by wt) starch.

with attached uronic acid. The very high proportion of xylose, relative to arabinose, in the insoluble fiber of oat bran from Sang could indicate a xylan as found in the hull and supports the hypothesis that the experimental milling process used (Salomonsson et al, 1984) did not completely remove the hull (Table 12).

The proportions of neutral sugar in the small amount of soluble fiber in the oat hull suggest that some cellulose may be lost with a bleaching step (Table 12). The major polymers in the insoluble fraction of the hull, indicated by the neutral sugar composition, are cellulose and a xylan.

COMPOSITION OF SOLUBLE AND INSOLUBLE DIETARY FIBER IN OTHER GRAINS AND PRODUCTS

One distinguishing characteristic of the fiber of oat groat and bran, compared to fiber from most other sources, is its β-glucan content (Tables 9 and 13). Barley is the only other fiber source with comparable amounts of β-glucan; rye has 1.7–2.9% β-glucan, but the remaining common cereals have <1% (Table 9). Many sources, however, have more total dietary fiber than oat groats or bran (Table 13). Barley meal and bran are higher in fiber than their oat counterparts. Processing barley meal into a flaked cereal shifts some of the insoluble fiber to the soluble fraction, and the neutral sugars in the insoluble fiber are decreased by pearling, a part of that processing (Marlett, 1991).

Rye meal and whole wheat flour have about the same amount of total fiber as the oat groat, but brans of both grains, which are not anatomically equivalent to oat bran, contain more fiber than oat bran (Tables 10 and 13). Wheat germ has about the same concentration of total fiber as oat bran. Corn bran is much higher in fiber than oat bran. The difference in the two corn bran samples in Table 13 may be that one was wet milled and the other dry milled. (The milling of neither sample was characterized.)

A second distinguishing feature of oat groats and bran, compared to other grain sources of fiber, is their relatively greater proportion of soluble fiber, ≈40–50% and ≈35–50%, respectively (Table 10). The soluble fiber in barley meal and bran from the Alva cultivar grown in Sweden was only 13 and 6%, respectively, of the total fiber (Table 13). Other barley cultivars that we have analyzed that were unprocessed or were processed into ready-to-eat cereals contained a much higher proportion of the total fiber in the soluble fraction (Table 13). It is possible that the Alva cultivar bran contained substantial amounts of husk. Rye and wheat products also contain less soluble fiber than oat groats or bran. Processed oat hull, perhaps similar to wood pulp cellulose, like corn and wheat brans, has a very high proportion of insoluble fiber. The total fiber concentration in oat hull is much higher than that in wheat bran but not necessarily than that in corn bran, the fiber content of which can vary with the milling process (Tables 10 and 13).

Neutral Sugar Composition of Dietary Fiber Fractions

All grain fibers, including oats, have more pentosan than cellulose in the insoluble fiber fraction (Tables 12 and 14). Using the β-glucan data for oatmeal and oat bran (Shinnick et al, 1988) and for unprocessed barley (Marlett, 1991) (Tables 10 and 13), the proportions of glucose (Tables 12 and 14) that reflect cellulose in the insoluble fiber fractions are oatmeal 1.28%, oat bran 2.25%, and barley 6.43% (dry weight). These data suggest that whole hulled barley contains more cellulose than either oat groat or bran. Similarly, wheat and corn bran contain more cellulose than oat bran (wheat bran for 1982, 9.2%; wheat bran for 1988, 12.4%; and corn bran, 14.8%). The glucose content indicates that the insoluble fraction of pea fiber is primarily cellulose and that the insoluble fraction of citrus fiber also has a substantial amount of cellulose (Table 14).

Pentosans are also the major hemicellulosic polysaccharides in the soluble fractions of grains other than oats and barley. These are mainly arabinoxylans (Table 14). As in oats, more than half of the neutral sugar content from the soluble fiber of barley is glucose from the β-glucan (Tables 12 and 14). Monosaccharide constituents of fiber from psyllium seed husk and polysaccharide fiber from soy show significant differences from cereals.

GRAVIMETRIC ANALYSIS OF DIETARY
FIBER FRACTIONS OF OATS

The dietary fiber of oat fractions has been analyzed gravimetrically by the AOAC method (Prosky et al, 1988), the method of Asp et al (1983), the method of Mongeau and Brassard (1989), and the NDF method of Robertson and Van Soest (1981) (Table 15). The NDF method underestimates the total dietary fiber in groats by >50%. The data from the Mongeau method, which uses the NDF procedure to measure insoluble fiber, indicate that the NDF content of the oat groat equals the insoluble fiber content. In general, the Asp and AOAC methods measure similar or lower amounts of total, soluble, and insoluble fibers in oat groats, compared to the chemical methods, with the exception of the Englyst method (Table 15 vs. Table 10). As always, some of the differences among the analyses of oat groats in Tables 15 and 10 could reflect sample differences.

TABLE 13
Composition (% of dry weight) of Soluble and Insoluble Dietary Fiber in Other Grains and Products

	Soluble			Insoluble				Total Fiber	Data from[a]
	Neutral Sugars	Uronic Acids	Total	Neutral Sugars	Uronic Acids	Klason Lignin	Total		
Barley meal, unprocessed									
Alva	2.1	UA[b]	NC[c]	11.3	0.2[d]	2.8	NC	16.4	H
Westbred Waxbar	1.2	0.0	4.9 (3.7)[e]	8.5	0.3	0.6	10.8 (1.4)	15.7	D
Barley meal, processed									
Westbred Waxbar[f]	1.3	0.1	6.3 (4.9)	4.5	0.1	0.9	6.0 (0.5)	12.3	D
	1.4	0.1	6.1 (4.6)	4.0	0.1	1.8	6.4 (0.5)	12.5	D
Sirobar, hulled[f]	2.4	0.2	7.1 (4.5)	3.7	0.1	1.2	5.3 (0.3)	12.4	D
Barley bran, Alva	1.8	UA	NC	20.3	0.9[d]	5.5	NC	28.5	H
Cellulose, wood pulp	0.6	Tr[g]	0.6	87.8	2.1	0	89.9	90.5	C
Citrus fiber	1.9	8.3	10.2	24.1	22.8	1.1	48.0	58.2	F
Corn bran	0.2	0.1	0.3	54.9	4.6	3.0	62.5	62.8	F
	0.9	0.3	1.2	76.0	5.7	2.3	84.0	85.2	A
Pea fiber	1.3	1.3	2.6	66.7	13.5	0	80.2	82.8	F
Psyllium seed husk	77.5	6.0	83.5[h]	16.0	0.6	0.4	17.0[h]	100.5	F
Rye meal	2.3	UA	NC	8.8	0.5[d]	1.9	NC	13.5	H
Rye bran	2.5	UA	NC	16.6	0.7[d]	2.9	NC	22.7	H
Soy polysaccharide	3.7	1.1	4.8	59.4	11.7	2.3	73.4	78.2	I

Wheat

White flour	1.8	T[i]	1.8	1.8	T	NA[j]	1.8	3.6	B
Whole wheat flour	1.0	0.1	1.1	2.0	0	0.2	2.2	3.3	D
Bran	2.7	0.1	2.8	7.9	0.2	NA	8.1	10.9	B
AACC wheat bran	2.9	0.3	3.2	37.5	0.9	NA	38.4	41.6	B
1982	1.5	0.2	1.7	29.9	1.9	5.9	37.7	39.4	G
1988	1.3	0.1	1.4	41.3	1.4	3.5	46.2	47.6	F
Germ	1.0	0.1	1.1	11.5	0.8	1.3	13.6	14.7	E

[a] A, Anderson and Bridges (1988); B, Cummings and Englyst (1987); C, Lopez-Guisa et al (1988); D, Marlett (1991); E, Marlett (1992); F, Marlett (unpublished); G, Neilson and Marlett (1983); H, Salomonsson et al (1984); I, Shinnick et al (1989).

[b] Unavailable, included in insoluble fiber uronic acids.

[c] Not calculable, as uronic acid is total value and does not reflect concentration in the fraction.

[d] Includes uronic acids from soluble fiber fraction.

[e] β-Glucan content, determined by enzymatic analysis, is in parentheses and included in the total.

[f] Processed into a flaked cereal.

[g] Trace, <0.05 g/100g.

[h] Soluble and insoluble fractions are separated on the basis of solubility in tertiary-butyl alcohol, not water, which is the standard method for determining soluble vs. insoluble fiber.

[i] Trace, not specified by authors.

[j] Not analyzed.

Two data sets on oat groats and one on oat bran are from collaborative studies; one sample was analyzed to evaluate the soluble-insoluble fiber analysis modification of the AOAC procedure (Prosky et al, 1988), and four samples each of groat and bran were analyzed by the AACC Oat Bran Committee to aid them in developing a definition (see Chapter 1) of oat bran (Table 15). There were 13 collaborators in the AOAC evaluation and 8–11, depending on the particular analysis, in the AACC Oat Bran Committee study. The latter group used the enzymatic method of McCleary and Glennie-Holmes (1985) as modified by L. Zygmunt (Quaker Oats Co.) for total β-glucan content. This method recently was adopted for first action by the official methods board of the AOAC International (Helrich, 1993).

The coefficient of variation for total fiber in oat groats from the AOAC trial data, a measure of variation among laboratories, was 17%; an average of the coefficients of variation of the four sets of analyses from the study organized by the AACC Oat Bran Committee was 8% (Table 15). Both collaborative studies experienced substantial variability in the soluble fiber determinations, obtaining coefficients of variation of 26–68%. The oat groat insoluble fiber data had a lower coefficient of variation in the AOAC collaborative study, 17%, than in the AACC studies, where the four data sets ranged from 15 to 38%.

The high value for oat bran fiber found by Krishnan et al (1987) using the NDF method without any treatment to remove starch can be accounted for by starch contamination of the fiber residue. The other oat bran fiber determined by the NDF approach (Sosulski and Wu, 1988) was of wild oats. However, the authors noted that the filtration step was very slow, and it is generally agreed that the chances of obtaining an inflated NDF value are much greater when filtration is slow (Robertson and Van Soest, 1981), which may account for the somewhat higher value in this analysis. With the exception of two oat bran samples, the total fiber contents of most samples obtained by gravimetric analysis are in general agreement with those obtained by chemical analysis (Table 15 vs. Table 10). The basis for the higher fiber content of the commercial hot cereal, Mother's Oat Bran, is not known; as noted earlier, the data do not agree with label values. The low concentrations of dietary fiber and ß-glucan in one of the oat bran samples of the AACC Oat Bran Committee study may be a consequence of either varietal differences or the milling technique used to obtain the bran. Regardless of the reason, it does not meet the analytical requirements of the AACC definition of oat bran (see Chapter 1).

As in the chemical analyses, the soluble fiber content of oat bran determined gravimetrically was a significant proportion of the total

TABLE 14
Neutral Sugar Composition of Soluble and Insoluble Dietary Fiber from Other Grains and Products

	Soluble, % of neutral sugars						Insoluble, % of neutral sugars						Data from[b]
	Glc[a]	Xyl	Ara	Man	Gal	Rha	Glc	Xyl	Ara	Man	Gal	Rha	
Barley meal, unprocessed													
Alva	53	23	19	1	4	Tr[c]	42	38	15	3	2	Tr	F
Westbred Waxbar	85	7	4	2	2	ND[d]	35	35	23	5	2	ND	C
Barley meal, processed[e]													
Westbred Waxbar	82	10	6	1	1	ND	31	37	24	6	2	ND	C
Australian hulled	80	11	6	2	1	ND	33	37	23	7	0	0	C
Barley bran	79	12	7	2	0	0	31	36	24	9	0	0	C
Oat bran	58	21	14	3	4	Tr	44	40	13	2	1	Tr	F
Cellulose, wood pulp	35	62	0	4	0	0	96	3	0	1	0	0	B
Citrus fiber	14	6	51	0	29	ND	53	9	21	5	12	ND	D
Corn bran	18	20	26	8	28	ND	27	41	23	1	8	ND	D
Pea fiber	10	36	38	0	16	ND	77	14	7	0	2	ND	D
Psyllium seed husk[f]	8	64	25	3	0	ND	23	15	36	14	12	ND	D
Rye meal	27	41	24	3	5	Tr	29	40	23	4	4	Tr	F
Rye bran	11	54	29	2	4	Tr	31	42	21	3	3	Tr	F
Soy polysaccharide	11	4	20	9	35	1[g]	14	11	22	2	43	1[g]	G
Wheat													
White flour	17	39	33	T[h]	11	T	17	44	33	6	T	T	A
Whole wheat flour	11	48	30	4	7	T	28	42	29	T	1	T	A
Bran	7	55	34	T	3	T	29	46	23	T	2	T	A
AACC wheat bran													
1982	20	54	26	<1	0	ND	31	46	23	<1	0	ND	E
1988	32	36	23	2	7	ND	30	44	25	1	0	ND	D

[a] Glc = glucose, Xyl = xylose, Ara = arabinose, Man = mannose, Gal = galactose, Rha = rhamnose.
[b] A, Cummings and Englyst (1987); B, Lopez-Guisa et al (1988); C, Marlett (1991); D, Marlett (unpublished); E, Neilson and Marlett (1983); F, Salomonsson et al (1984); G, Shinnick et al (1989).
[c] Trace, <0.05 g/100 g.
[d] Not detected; Rha and Gal coelute by the high-performance liquid chromatographic method used.
[e] Processed into a flaked cereal.
[f] Soluble and insoluble fractions are separated on the basis of solubility in tertiary-butyl alcohol, not water.
[g] Remainder of neutral sugars is fucose.
[h] Trace, concentration not specified.

(Table 10 vs. Table 15). The total ß-glucan concentrations measured enzymatically are similar to those obtained by measuring the soluble and insoluble fractions separately by the method of Shinnick et al (1988), which used the difference in glucose content between a sample treated with ß-glucanase and an untreated sample.

TABLE 15
Dietary Fiber (Gravimetric) and β-Glucan Analysis of Oats (g/100 g dry wt)

	Method of Analysis[a]	Dietary Fiber			Total β-Glucan[b]	Data From[c]
		Total	Soluble	Insoluble		
Groat						
Oatmeal	NDF	5.0	NM[d]	NM	NM	E
	Mongeau	10.6	5.2	5.4	INC[e]	H
	AOAC	11.8 ± 2.0[f] (17)[g]	4.2 ± 1.1[f] (26)	6.3 ± 1.1[f] (17)	INC	K
	AOAC	9.5 ± 0.8[h]	3.2 ± 1.5[h]	6.9 ± 1.5[h]	4.1 ± 0.4[h]	A
		10.6 ± 1.7[h]	3.8 ± 2.6[h]	6.3 ± 2.2[h]	4.1 ± 0.4[h]	A
		8.5 ± 0.9[h]	2.2 ± 1.1[h]	5.6 ± 1.6[h]	3.0 ± 0.4[h]	A
		9.3 ± 1.1[h]	2.4 ± 1.2[h]	5.8 ± 1.9[h]	3.1 ± 0.3[h]	A
Selma	Asp	11.5	2.7	8.8	INC	C
Bran						
Selma	Asp	16.4	6.2	10.2	INC	J
Mother's	Asp	25.6	9.5	16.1	INC	J
NS[i]	NDF	31.2	NM	NM	NM	F
Wild	NDF/ AOAC	21.3/20.0	NM	NM	NM	L
NS	AOAC	18.2 ± 1.1	7.1 ± 0.6	11.8 ± 1.9	7.2 ± 0.5	A
		17.7 ± 0.7	6.5 ± 0.3	11.3 ± 2.3	6.6 ± 0.4	A
		18.6 ± 1.0	6.4 ± 0.5	12.7 ± 1.6	6.6 ± 0.4	A
		12.0 ± 1.1	3.6 ± 0.5	8.3 ± 0.5	4.1 ± 0.4	A
Hull						
Selma	Asp	83.9	0.3	83.6	NM	J
NS	NDF	70.0	NM	NM	NM	I
NS	NDF	75.8	NM	NM	NM	D
Processed[j]	AOAC	81.5	NA[k]	NA	INC	B
Bleached	AOAC	95.6	NA	NA	INC	B
Processed[j]	NDF	71.3	NM	NM	NM	G
Bleached	NDF	84.4	NM	NM	NM	G

[a]For total, soluble, and insoluble fiber data. NDF = neutral detergent fiber.
[b]Determined by the method of McCleary and Glennie Holmes (1985).
[c]A, AACC Oat Bran Committee (1989, *unpublished*) (four samples analyzed by eight to 11 laboratories); B, Dougherty et al (1988); C, Frølich and Nyman (1988); D, Garleb et al (1988); E, Johnson and Marlett (1986); F, Krishnan et al (1987); G, Lopez-Guisa et al (1988); H, Mongeau et al (1989); I, Moore et al (1986); J, Nyman and Asp (1988); K, Prosky et al (1988); L, Sosulski and Wu (1988).
[d]Not measured by method.
[e]Not analyzed separately but included in soluble and insoluble fiber values.
[f]Mean ± standard deviation of analyses from 13 laboratories.
[g]Coefficient of variation.
[h]Mean ± standard deviation of analyses from 8–11 laboratories.
[i]Not specified.
[j]Coated with 10% (by wt) starch.
[k]Not analyzed.

Gravimetric analysis, like chemical methods, shows that oat hull is a concentrated source of insoluble fiber; however, the NDF method yielded lower values than either the AOAC or Asp procedures (Table 15).

EFFECTS OF PROCESSING ON DIETARY FIBER FROM OAT GROATS AND BRAN

Commercial processing can affect the proportion of total oat fiber that is extracted into the soluble fraction (Table 16). To evaluate the effect of commercial processing, two batches of high-fiber oat flours were prepared by two separate millings of oat bran and then were processed by low-pressure extrusion to a dough, which was formed into sweetened flakes (Shinnick et al, 1988). The proportion of the total fiber extracted as soluble fiber increased with processing of Batch I high-fiber oat flour, both because the neutral sugar content (but not ß-glucan) of the soluble fiber fraction increased and because Klason lignin decreased. With processing of Batch II, the soluble fraction increased because the concentrations both of neutral sugars from the soluble dietary fiber and of soluble ß-glucan increased, while Klason lignin in the insoluble fiber fraction decreased.

TABLE 16
Effect of Processing on Dietary Fiber Content
from Oat Groats and Bran[a]

	Content (% of total fiber)					Total (% dry wt of sample)	
	Neutral Sugars	ß-Glucans	Uronic Acids	Klason Lignin	Total	In Fraction	Dietary Fiber
Oat bran							
Soluble[b]	9	29	1	0	39	7.2	
Insoluble	27	12	2	20	61	11.4	18.6
High-fiber oat flour, batch I							
Unprocessed							
Soluble	5	37	1	0	43	10.6	
Insoluble	26	10	2	19	57	14.3	24.9
Processed							
Soluble	15	35	1	0	51	8.3	
Insoluble	28	8	2	11	49	8.0	16.3
High-fiber oat flour, batch II							
Unprocessed							
Soluble	2	29	1	0	32	8.2	
Insoluble	22	24	2	20	68	16.9	25.0
Processed							
Soluble	9	41	1	0	51	9.1	
Insoluble	25	10	2	13	50	9.0	18.0

[a]Recalculated from data of Shinnick et al (1988).
[b]Soluble and insoluble fiber fractions.

Conclusions

The substantial emphasis on the potential health benefits of consuming oat bran has led to the availability of a wide variety of oat fiber sources in the marketplace. In March 1989, the American Association of Cereal Chemists convened a committee to develop a definition for oat bran. The definition that was presented and approved by the AACC Committee on Oat Bran in Washington, DC, October 31, 1989, is reported in Chapter 1. Most, but not all, of the oat bran data reviewed in this chapter met this definition.

Comparisons of the composition of oat groats and bran with the composition of other grains clearly indicate that these fractions of oat grain are good sources of a high-quality protein and may provide nutritionally significant amounts of calcium, iron, and zinc. In addition, although they are not the most concentrated sources of grain fibers, the oat groat and bran contain unique mixtures of dietary fiber. They are higher in soluble fiber than most foods. The average mixed diet contains about 25–30% of the total fiber as soluble fiber (Pilch, 1987); about 35–50% of the fiber in oat groats and bran is soluble. Further, these two fractions of the oat grain are the major source of one type of fiber, $(1\rightarrow3),(1\rightarrow4)$-ß-D-glucan. In contrast, the oat hull, although a much more concentrated source of fiber, is >95% insoluble fiber and contains no measurable amounts of ß-glucan.

ACKNOWLEDGMENTS

The author appreciated the advice and direction given by David M. Peterson, University of Wisconsin-Madison and Peter J. Wood, Agriculture Canada, during the preparation of this chapter.

LITERATURE CITED

AMAN, P., and HESSELMAN K. 1984. Analysis of starch and other main constituents of cereal grains. Swed. J. Agric. Res. 14:135-139.

AMAN, P., and HESSELMAN, K. 1985. An enzymic method for analysis of total mixed-linkage ß-glucans in cereal grains. J. Cereal Sci. 3:251-257.

ANDERSON, J. W., and BRIDGES, S. R. 1988. Dietary fiber content of selected foods. Am. J. Clin. Nutr. 47:440-447.

ASP, N. G., JOHANSSON, C. G., HALLMER, H., and SILJESTROM, M. 1983. Rapid enzymatic assay of insoluble and soluble dietary fiber. J. Agric. Food Chem. 31:476-482.

BINGHAM, S. 1987. Definitions and intakes of dietary fiber. Am. J. Clin. Nutr. 45:1226-1231.

CARR, J. M., GLATTER, S., JERACI, J. L., and LEWIS, B. A. 1990. Enzymic determination of ß-glucan in cereal-based food products. Cereal Chem. 67:226-229.

CHEN, H., RUBENTHALER, G. L., LEUNG, H. K., and BARANOWSKI, J. D. 1988. Chemical, physical, and baking properties of apple fiber compared with wheat and oat bran. Cereal Chem. 65:244-247.

CUMMINGS, J. H., and ENGLYST, H. N. 1987. The development of methods for the measurement of dietary fibre in food. Pages 188-220 in: Cereals in a European Context. First European Congress of Food Science and Technology. I. D. Morton, ed. VCH Publishers, New York.

DEANE, D., and COMMERS, E. 1986. Oat cleaning and processing. Pages 371-412 in: Oats: Chemistry and Technology. F. H. Webster, ed. Am. Assoc. Cereal Chem., St. Paul, MN.

DOUGHERTY, M., SOMBKE, R., IRVINE, J., and RAO, C. S. 1988. Oat fibers in low calorie breads, soft-type cookies, and pasta. Cereal Foods World 33:424-427.

DRAPER, S. R. 1973. Amino acid profiles of chemical and anatomical fractions of oat grains. J. Sci. Food Agric. 24:1241-1250.

ENGLYST, H. N., and CUMMINGS, J. H. 1982. Determination of non-starch polysaccharides in plant foods by gas-liquid chromatography of constituent sugars as alditol acetates. Analyst 107:307-318.

FRØLICH, W., and NYMAN, M. 1988. Minerals, phytate and dietary fibre in different fractions of oat-grain. J. Cereal Sci. 7:73-82.

GARLEB, K. A., FAHEY, G. C., LEWIS, S. M., KERLEY, M. S., and MONTGOMERY, L. 1988. Chemical composition and digestibility of fiber fractions of certain by-product feedstuffs fed to ruminants. J. Anim. Sci. 66:2650-2662.

GRAHAM, H., RYDBERG, M.-B. G., and AMAN, P. 1988. Extraction of soluble dietary fiber. J. Agric. Food Chem. 36:494-497.

HELRICH, K. 1993. (1→3)(1→4)-β-D-Glucan in oat and barley fractions and ready-to-eat cereals, enzymatic-spectophotometric method. Official Methods of Analysis, 15th ed., 4th suppl. AOAC Interntional, Arlington, VA. In press.

HENRY, R. J. 1985. A comparison of the non-starch carbohydrates in cereal grains. J. Sci. Food Agric. 36:1243-1253.

JOHNSON, E. J., and MARLETT, J. A. 1986. A simple method to estimate neutral detergent fiber content of typical daily menus. Am. J. Clin. Nutr. 44:127-134.

KRISHNAN, P. G., CHANG, K. C., and BROWN, G. 1987. Effect of commercial oat bran on the characteristics and composition of bread. Cereal Chem. 64:55-58.

LARM, D., THEANDER, O., and AMAN P. 1975. Structural studies on a water-soluble arabinan isolated from rapeseed (Brassica napus). Acta Chem. Scand. B29:1011-1014.

LOPEZ-GUISA, J. M., HARNED, M. C., DUBIELZIG, R., RAO, S. C., and MARLETT, J. A. 1988. Processed oat hulls as potential dietary fiber sources. J. Nutr. 118:953-962.

MA, C.-Y. 1983. Chemical characterization and functionality assessment of protein concentrates from oats. Cereal Chem. 60:36-42.

MacARTHUR, L. A., and D'APPOLONIA, B. L. 1979a. Comparison of oat and wheat carbohydrates. I. Sugars. Cereal Chem. 56:455-457.

MacARTHUR, L. A., and D'APPOLONIA, B. L. 1979b. Comparison of oat and wheat carbohydrates. II. Starch. Cereal Chem. 56:458-461.

MacARTHUR-GRANT, L. A. 1986. Sugars and non-starchy polysaccharides in oats. Pages 75-91 in: Oats: Chemistry and Technology. F. H. Webster, ed. Am. Assoc. Cereal Chem., St. Paul, MN.

MARLETT, J. A. 1990a. Analysis of dietary fiber in human foods. Pages 31-48 in: Dietary Fiber: Chemistry, Physiology and Health Effects. D. Kritchevsky, C.

Bonfield, and J. W. Anderson, eds. Plenum Press, New York.

MARLETT, J. A. 1990b. Issues in dietary fiber analysis. Pages 183-192 in: Symposium on Dietary Fiber—New Developments: Physiological Effects and Physiochemical Properties. I. Furda and C. Brine, eds. Plenum Press, New York.

MARLETT, J. A. 1991. Dietary fiber content and effect of processing on two barley varieties. Cereal Foods World 36:576-578.

MARLETT, J. A. 1992. The content and composition of dietary fiber in 117 frequently consumed foods. J. Am. Diet. Assoc. 92:175-186.

MARLETT, J. A., and NAVIS, D. 1988. Comparison of gravimetric and chemical analyses of total dietary fiber in human foods. J. Agric. Food Chem. 36:311-315.

MARLETT, J. A., CHESTER, J. G., LONGACRE, M. J., and BOGDANSKE, J. J. 1989. Recovery of soluble dietary fiber is dependent on the method of analysis. Am. J. Clin. Nutr. 50:479-485.

MATZ, S. A. 1969. Oats. Pages 78-96 in: Cereal Science. S. A. Matz, ed. AVI Publishing Company, Westport, CT.

McCLEARY, B. V., and GLENNIE-HOLMES, M. 1985. Enzymic quantification of (1→3), (1→4)-ß-D-glucan in barley and malt. J. Inst. Brew. 91:285-295.

McKECHNIE, R. 1983. Oat products in bakery foods. Cereal Foods World 28:635-637.

MONGEAU, R., and BRASSARD, R. 1986. A rapid method for the determination of soluble and insoluble dietary fiber: Comparison with AOAC total dietary fiber procedure and Englyst's method. J. Food Sci. 51:1333-1336.

MONGEAU, R., and BRASSARD, R. 1989. A comparison of three methods for analyzing dietary fiber in 38 foods. J. Food Compos. Anal. 2:189-199.

MONGEAU, R., BRASSARD, R., and VERDIER, P. 1989. Measurement of dietary fiber in a total diet study. J. Food Compos. Anal. 2:317-326.

MOORE, R. J., KORNEGAY, E. T., and LINDEMANN, M. D. 1986. Effect of dietary oat hulls or wheat bran on mineral utilization in growing pigs fed diets with or without salinomycin. Can. J. Anim. Sci. 66:267-276.

MOREY, D. D., and EVANS, J. J. 1983. Amino acid composition of six grains and winter wheat forage. Cereal Chem. 60:461-464.

MORGAN, D. E. 1968. Note on variations in the mineral composition of oat and barley grown in Wales. J. Sci. Food Agric. 19:393-395.

MOSSE, J., HUET, J.-C., and BAUDET, J. 1988a. The amino acid composition of triticale grain as a function of nitrogen content: Comparison with wheat and rye. J. Cereal Sci. 7:49-60.

MOSSE, J., HUET, J.-C., and BAUDET, J. 1988b. The amino acid composition of rice grain as a function of nitrogen content as compared with other cereals: A reappraisal of rice chemical scores. J. Cereal Sci. 8:165-175.

NEILSON, M. J., and MARLETT, J. A. 1983. A comparison between detergent and non-detergent analyses of dietary fiber in human foodstuffs, using high performance liquid chromatography to measure neutral sugar composition. J. Agric. Food Chem. 31:1342-1347.

NEWMAN, R. K. and NEWMAN, C. W. 1991. Barley as a food grain. Cereal Foods World 36:800-805.

NEWMARK, H. L. 1987. Plant phenolics as inhibitors of mutational and pre-carcinogenic events. Can. J. Physiol. Pharmacol. 65:461-466.

NYMAN, M. G.-L., and ASP, N.-G. L. 1988. Fermentation of oat fiber in the rat intestinal tract: A study of different cellular areas. Am. J. Clin. Nutr. 48:274-278.

ORR, M. L., and WATT, B. K. 1968. Amino acid content of foods. Pages 28-29 in: Home Economic Research Report no. 4. United States Department of Agriculture, Washington, DC.

PETERSON, D. M. 1992. Composition and nutritional characteristics of oat grain and oat products. Pages 265-292 in: Oat Science and Technology. H. G. Marshall and M. E. Sorrells, eds. American Society of Agronomy and Crop Science Society of America, Madison, WI.

PILCH, S. M., ed. 1987. Physiological Effects and Health Consequences of Dietary Fiber. Life Sciences Research Office of the Federation of American Societies for Experimental Biology, Bethesda, MD.

POMERANZ, Y., and DIKEMAN, E. 1983. Minerals and protein contents in hard red winter wheat flours. Cereal Chem. 60:80-82.

POMERANZ, Y., YOUNGS, V. L., and ROBBINS, G. S. 1973. Protein content and amino acid composition of oat species and tissues. Cereal Chem. 50:702-707.

PRENTICE, N., BABLER, S., and FABER, S. 1980. Enzymic analysis of ß-D-glucans in cereal grains. Cereal Chem. 57:198-202.

PROSKY, L., ASP, N.-G., FURDA, I., DEVRIES, J. W., SCHWIEZER, T. F., and HARLAND, B. F. 1985. Determination of dietary fiber in foods and food products: Collaborative study. J. Assoc. Off. Anal. Chem. 68:677-679.

PROSKY, L., ASP, N. G., SCHWIEZER, T. F., DEVRIES, J. W., and FURDA, I. 1988. Determination of insoluble, soluble and total dietary fiber in foods and food products: Interlaboratory study. J. Assoc. Off. Anal. Chem. 71:1017-1023.

PROSKY, L., ASP, N. G., SCHWEIZER, T. F., DEVRIES J. W., and FURDA, I. 1992. Determination of Insoluble and Soluble Dietary Fiber in Food and Food Products: Collaborative study. J. Assoc. Off. Anal. Chem. 75:360-367.

PROSKY, L. ASP, N. G., SCHWEIZER, T. F., DEVRIES, J. W., FURDA, I., and LEE, S. C. 1993. Determination of soluble dietary fiber in food and food products: Collaborative study. J. AOAC Int. In press.

ROBERTSON, J. B., and VAN SOEST, P. J. 1981. The detergent system of analysis and its application to human foods. Pages 123-158 in: The Analysis of Dietary Fiber in Food. W. P. T. James and O. Theander, eds. Marcel Dekker, New York.

SALOMONSSON, A. C., THEANDER, O., and WESTERLUND, E. 1984. Chemical characterization of some Swedish cereal whole meal and bran fractions. Swed. J. Agric. Res. 14:111-117.

SANCHEZ-MARROQUIN A., DELVALLE, F. R., ESCOBEDO, M., AVITIA, R., MAYA, S., and VEGA, M. 1986. Evaluation of whole amaranth (Amaranthus cruentus) flour, its air-classified fractions, and blends of these with wheats and oats as possible components for infant formulas. J. Food Sci. 51:1231-1234, 1238.

SHINNICK, F. L., LONGACRE, M. J., INK, S. L., and MARLETT, J. A. 1988. Oat fiber: Composition vs. physiological function. J. Nutr. 118:144-151.

SHINNICK, F. L., HESS, R. L., FISCHER, M. H., and MARLETT, J. A. 1989. Apparent nutrient absorption and upper gastrointestinal transit with fiber-containing enteral feedings. Am. J. Clin. Nutr. 49:471-475.

SIDDIQUI, I. R., and WOOD, P. J. 1974. Structural investigation of oxalate-soluble rapeseed (Brassica campestris) polysaccharides. Part III. An arabinan. Carbohydr. Res. 36:35-44.

SOSULSKI, F. W., and WU, K. K. 1988. High-fiber breads containing field pea hulls, wheat, corn, and wild oat brans. Cereal Chem. 65:186-191.

SOUTHGATE, D. A. T. 1969. Determination of carbohydrates in foods. II.

Unavailable carbohydrates. J. Sci. Food Agric. 20:332-335.

SOUTHGATE, D. A. T. 1981. Use of the Southgate method for unavailable carbohydrates in the measurement of dietary fiber. Pages 1-19 in: The Analysis of Dietary Fiber in Food. W. P. T. James and O. Theander, eds. Marcel Dekker, New York.

SUMNER, A. K., GEBRE-EGZIABHER, A., TYLER, R. T., and ROSSNAGAEL, B. G. 1985. Composition and properties of pearled and fines fractions from hulled and hull-less barley. Cereal Chem. 62:112-116.

THEANDER, O., and AMAN, P. 1979. Studies on dietary fibers. 1. Analysis and chemical characterization of water-soluble and water-insoluble dietary fibres. Swed. J. Agric. Res. 9:97-106.

THEANDER, O., and AMAN, P. 1981. Analysis of dietary fibers and their main constituents. Pages 51-70 in: The Analysis of Dietary Fiber in Food. W. P. T. James and O. Theander, eds. Marcel Dekker, New York.

THEANDER, O., and WESTERLUND, E. 1986. Studies on dietary fiber. 3. Improved procedures for analysis of dietary fiber. J. Agric. Food Chem. 34:330-336.

THOMKE, S. 1988. Oats as an animal feed. Pages 164-175 in: Proc. Int. Oat Conference, 3rd. B. Mattsson and R. Lyhagen, eds. Svalof AB, Sweden.

VAN SOEST, P. J., and WINE, R. H. 1963. Use of detergents in the analysis of fibrous foods. IV. Determination of plant cell-wall constituents. J. Assoc. Off. Anal. Chem. 50:50-55.

WEAVER, S. H. 1988. The history of oat milling. Pages 47-50 in: Proc. Int. Oat Conference, 3rd. B. Mattsson and R. Lyhagen, eds. Svalof AB, Sweden.

WEBSTER, F. H., ed. 1986a. Oats: Chemistry and Technology. Am. Assoc. Cereal Chem., St. Paul, MN.

WEBSTER, F. H. 1986b. Oat utilization: Past, present, and future. Pages 413-426 in: Oats: Chemistry and Technology. F. H. Webster, ed. Am. Assoc. Cereal Chem., St. Paul, MN.

WELCH, R. W., and LLOYD, J. D. 1989. Kernel $(1\rightarrow3)$ $(1\rightarrow4)$-ß-D-glucan content of oat genotypes. J. Cereal Sci. 9:35-40.

WOOD, P. J., WEISZ, J., and FEDEC, P. 1991. Potential for ß-glucan enrichment in brans derived from oat (Avena sativa L.) cultivars of different $(1\rightarrow3),1\rightarrow4)$-ß-D-glucan concentrations. Cereal Chem. 68:48-51.

YLI-HALLA, M., and PALKO, J. 1987. Mineral element content of oats (Avena sativa L.) in an acid sulphate soil area of Tupos village, northern Finland. J. Agric. Sci. Finland 59:73-78.

YOUNGS, V. L. 1974. Extraction of a high-protein layer from oat groat, bran and flour. J. Food Sci. 39:1045-1046.

YOUNGS, V. L. 1978. Oat lipids. Cereal Chem. 55:591-597.

YOUNGS, V. L., PETERSON, D. M., and BROWN C. M. 1982. Oats. Pages 49-105 in: Advances in Cereal Science and Technology, vol. 5. Y. Pomeranz, ed. Am. Assoc. Cereal Chem., St. Paul, MN.

ZARKADAS, C. G., HULAN, H. W., and PROUDFOOT, F. G. 1982. A comparison of the amino acid composition of two commercial oat groats. Cereal Chem. 59:323-327.

CHAPTER 4

Physicochemical Characteristics and Physiological Properties of Oat (1→3),(1→4)-ß-D-Glucan

Peter J. Wood
Centre for Food and Animal Research
Agriculture Canada
Ottawa, Ontario K1A 0C6, Canada

Introduction

Oats have long been associated with health and vitality. For example, a 19th century New England hiking club referred to themselves as the "Oatmeal Crusaders" (Waterman and Waterman, 1989), and the expression "feeling your oats" is part of our everyday language. This reputation is deserved since oats are a nutritious cereal (Lockhart and Hurt, 1986).

Until recently, processing and uses had not progressed much beyond the staple, rolled oats. However, the oat groat is an attractive potential source of protein, starch, and oil if suitable separation processes and marketable functional properties can be found. Commercial processes were devised in the 1970s with protein as the anticipated primary marketable product, and one such process was developed at the Food Research Centre in Ottawa (Paton, 1977). It was based on separation of protein and starch by extraction of the protein into aqueous sodium carbonate at pH 10. Wet processing of oats results in solubilization of the endospermic cell wall polysaccharide (1→3),(1→4)-ß-D-glucan, or ß-glucan. The viscosity of this polysaccharide causes difficulties in centrifugal or filtration separations but is sufficiently high that the crude extract of oat gum, which is mainly oat ß-glucan, was quickly recognized as a potentially valuable food hydrocolloid by-product from value-added processing (Hohner and

Hyldon, 1977; Wood et al, 1977, 1978).

Despite a theoretical value for fractionated oat products, their marketability and other economic factors were apparently not sufficiently attractive to encourage commercial development. However, coincidental to the waning interest in oats as a protein and starch source, epidemiologically based evidence that diets of Western developed societies were deficient in dietary fiber (Burkitt and Trowell, 1975) achieved public acceptance. Animal and clinical studies indicated that soluble, viscous, nonstarch polysaccharides, now referred to as soluble dietary fiber, had particular value in reducing serum choesterol concentrations (Kay and Truswell, 1980). Oat bran became recognized as a good source of such material through the pioneering studies of J. W. Anderson (Chapter 6).

Health benefits associated with oat bran consumption have been attributed to the presence of the ß-glucan, which is one and a half to two times higher in the bran than in the groat (Wood et al, 1991a). Because of this, interest revived in oat ß-glucan, the first cereal mixed-linkage ß-glucan to be structurally characterized in detail (Peat et al, 1957; Parrish et al, 1960). This followed a long period in which the ß-glucan of barley had received most attention because of its influence on the commercial value of feed and malt (Woodward and Fincher, 1983). Oat ß-glucan has been comprehensively reviewed and compared with barley ß-glucan (Wood, 1986). This chapter therefore deals only briefly with early literature and concentrates on recently published physicochemical and analytical data and physiological studies in which purified oat gum has been used.

Structure

Purified ß-glucan from oats is a linear, unbranched polysaccharide composed of 4-O-linked ß-D-glucopyranosyl units (≈70%) and 3-O-linked ß-D-glucopyranosyl units (≈30%) (Peat et al, 1957; Parrish et al, 1960; Clarke and Stone, 1966; Aspinall and Carpenter, 1984).

The evidence suggests that the (1→3) linkages occur singly (Dais and Perlin, 1982; Vårum and Smidsrød 1988), whereas most of the (1→4) linkages occur in groups of two or three (Perlin and Suzuki, 1962; Dais and Perlin, 1982; Aspinall and Carpenter, 1984). The resultant structure for oat ß-glucan is a polysaccharide composed mainly of ß-(1→3)-linked cellotriosyl and cellotetraosyl units. Oat and barley ß-glucan have been generally categorized as structurally similar, and both contain more of the (1→4)-linked cellotetraosyl units than lichenan, a (1→3),(1→4)-ß-D-glucan from the lichen *Cetraria islandica* (Parrish et al, 1960; Perlin and Suzuki, 1962; Fleming and Manners, 1966).

Methods of structural analysis developed for barley ß-glucan by Fincher and colleagues (Woodward and Fincher, 1982; Woodward et al, 1983) were recently applied to oat ß-glucan (Wood et al, 1991b). Methylation analysis was used to characterize oligosaccharide products produced by the action of a (1→3),(1→4)-ß-D-glucan-4-glucanohydrolase (lichenase), which cleaves the (1→4) linkage of the 3-O-substituted glucose units in the ß-glucan. The released oligosaccharides, which are ß-(1→4)-linked in the original chain, are the building blocks of the polysaccharide. The reducing end is identified by a second methylation of the prereduced oligosaccharide. The results, summarized in Table 1, confirmed that the major structural features were (1→3)-ß-linked cellotriosyl and cellotetraosyl units. As in barley ß-glucan, small amounts of ß-(1→3)-linked cello-oligosaccharides with a higher degree of polymerization (DP) were detected. Analysis of the insoluble material (6% by weight) produced by the enzyme indicated that the average number of consecutive (1→4) linkages was seven to eight, one to two units less than suggested for similar material from barley (Woodward et al, 1983).

Methylation analysis of the purified intact oat ß-glucan produced values that agreed with results of enzyme analysis. The percentage of (1→4) linkages was similar to that for barley ß-glucan and somewhat higher than for lichenan (Wood et al, 1989a).

^{13}C-nuclear magnetic resonanace (NMR) spectra for each ß-glucan were almost identical (Fig. 1). The single resonances from the 3-linked residue (e.g., 86.2 ppm for C-3) are interpreted to indicate a single environment, in terms of adjacent residues, for this carbon and suggest that most of the (1→3) linkages occur singly. This may be contrasted with the resonances from C-4 at 79.6–79.9 ppm, which indicate at least three environments for the C-4 involved in the (1→4) linkage. The

TABLE 1
Results of Methylation Analysis of Oligosaccharides
Produced by the Action of Lichenase
on the (1→3),(1→4)-β-D-Glucan of Oats[a]

Oligosaccharide	Glycosyl Residue,[b] molar proportions		
	T-Glc	3-Glc	4-Glc
Trisaccharide	1	1.03[c]	1.02
Tetrasaccharide	1	0.98[c]	1.97
Pentasaccharide	1	0.81[c]	2.66
Hexasaccharide	1	1.36[d]	3.42
Water-insoluble	1	0.80	6.19

[a]Adapted from Wood et al (1991b).
[b]T-Glc = terminal glucose, nonreducing end; 3-Glc = 3-linked glucose; 4-Glc = 4-linked glucose.
[c]Identified as a residue on reducing end, corrected for losses during methylation.
[d]Evidence for ~0.6-mol-proportion nonreducing 3-Glc.

basis for this can be readily seen on examining the structure represented in Figure 2. In view of the previously established distinctness of lichenan (Perlin and Suzuki, 1962), in which there were relatively few of the cellotetraosyl units, the closely similar C-4 signal, noted before by Dais and Perlin (1982), was somewhat surprising. If the higher field signal (79.6 ppm; incorrectly identified as 79.9 ppm in Wood et al, 1991b) represents C-4 of a 4-O-linked residue flanked on both the reducing and nonreducing ends by further 4-O-linked residues, then only those areas with three or more consecutive (1→4) linkages contribute to the signal. Unfortunately, resolution of this signal was poor, and quantitative differences were difficult to assess. Details of fine structure might be further evaluated by higher field spectrometers.

It is clear, however, that despite similar ^{13}C-NMR spectra, the three ß-glucans are structurally distinct. This was demonstrated by quantitative analysis by high-performance liquid chromatography (HPLC) of lichenase-released oligosaccharides (Wood et al, 1989b,

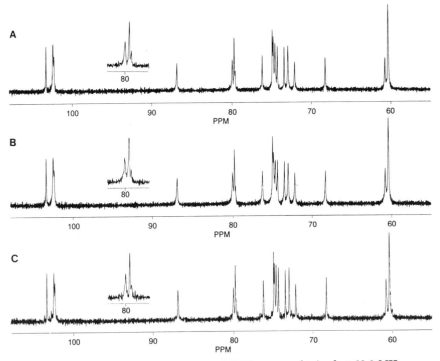

Figure 1. ^{13}C-Nuclear magnetic resonance (NMR) spectra obtained at 62.8 MHz on a Bruker WM250 NMR spectrometer. Samples (4%, w/v) in deuterated dimethyl-sulfoxide: A, oat ß-glucan; B, barley ß-glucan; C, lichenan. (Reprinted, with permission, from Wood et al, 1991b)

1991b). The difference between lichenan and oat or barley ß-glucans is striking (Fig. 3) and confirms the results of the early elegant studies of Perlin and Suzuki (1962).

Some care is required in generalizing from studies of purified material since observed differences, or similarities, may be a function of the method of extraction and purification (Wilkie, 1979) or of cultivar and environment, as much as of species. The HPLC method of analysis of lichenase-released oligosaccharides (Wood, 1985; Wood et al, 1991b) allows rapid structural evaluation without prior extraction and purification. Thus, whole groat flour, bran, crude oat gum, and derived purified ß-glucan showed identical ratios of tri- to tetrasaccharides (Wood et al, 1989a, 1991b). Consistent differences were observed between oats, in which approximately one third of the structure was ß-(1→3)-linked cellotetraosyl units, and barley and wheat, where the proportion was closer to one quarter (Table 2). The ratio range (2.5–3.2) for barley seemed more variable than for oats (1.9–2.2). Inclusion of other *Avena* species extended the range slightly to 1.8–2.3 (Miller et al, 1991). Evidently this structural characteristic is conserved. The average ratio for oat groats (2.1) was similar to that (2.2) reported by Yamamoto and Nevins (1978) for oat coleoptile.

The importance of structure lies in the influence this may have on physical and physiological properties. Differences in solubility and viscosity, described in the next section, may result in part from differences in fine structure.

The ß-glucans, like starch, are composed entirely of glucose, but the different structure of starch, in which the glucopyranose units are joined by α-(1→4) and (1→6) units, results in profoundly different properties. Cellulose, apparently more closely related, with all ß-(1→4) linked glucans, or curdlan with all ß-(1→3)-linked glucans, also possesses very different properties that can be related to differences in conformation, or shape.

No X-ray crystallographic studies have been reported for oat ß-glucan, but barley ß-glucan had a pattern similar to that of lichenan, which was characterized as a three-fold helix with three cello-

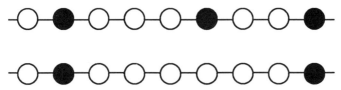

Figure 2. Schematic representation of oat ß-glucan structure, showing ß-(1→3)-linked cellotriosyl and cellotetraosyl units and a longer sequence of consecutive (1→4)-linked ß-D-glucopyranosyl units: ○, (1→4)-linked ß-D-glucopyranose; ●, (1→3)-linked ß-D-glucopyranose.

triose units per turn (Tvaroska et al, 1983). Therefore, increasing the proportion of cellotetraosyl units did not greatly influence conformation, although the barley ß-glucan showed less crystallinity than lichenan.

Conformational analysis has been used to predict molecular ex-

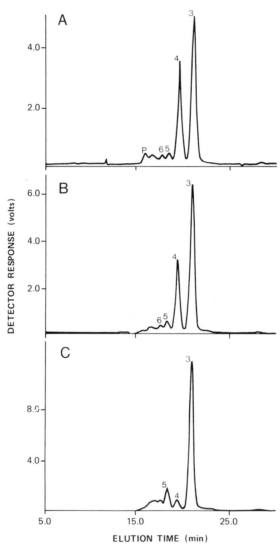

Figure 3. High-performance liquid chromatography on HPX-42A (0.4 ml/min) of oligosaccharides released by lichenase from oat gum (**A**), barley ß-glucan (**B**), and lichenan (**C**). Detection by the orcinol-sulfuric acid reaction. Peaks 3, 4, 5, and 6 are tri-, tetra-, penta-, and hexasaccharide, respectively. P = polymer peak. (Reprinted, with permission, from Wood et al, 1991b)

tension in solution (Buliga et al, 1986). The calculations were based
on a random arrangement of the (1→3)-linked cellotriosyl and
cellotetraosyl units within the same molecule, an assumption supported by analysis of the rates of oligosaccharide production by (1→3),
(1→4)-ß-D-glucan-4-glucanohydrolase from barley (Staudte et al,
1983). The calculations showed that the experimentally obtained
values for molecular dimensions could not be explained on the basis
of ß-(1→3)-linked cellotriosyl and cellotetraosyl units alone and that
sequences of four to 10 consecutive (1→4) linkages, although present
in low mole proportions (≈2%), would disproportionately influence
molecular dimension. A structure composed entirely of ß-(1→3)-
linked cellotriosyl and cellotetraosyl units underestimated the
molecular extension in solution. Although calculations were based on
the molar proportions of cellotriosyl and cellotetraosyl units found for
barley ß-glucan, it is evident that the small change in proportion in
oat ß-glucan would not greatly change calculated molecular extension. However, small proportions of longer sequences of (1→4)
linkages do significantly increase, and sequences of more than one
consecutive (1→3) linkage significantly decrease, theoretical molecular dimensions of ß-glucans of otherwise similar DP and structure.
These features would therefore be expected to exert a major influence on
viscosity. Furthermore, it has been suggested that differences in
amount of consecutive (1→4) linkages might explain solubility differences of some cereal ß-glucan fractions (Beresford and Stone, 1983;
McCleary, 1988; Woodward et al, 1988).

It is therefore evident that despite some differences in the now
well-established major structural features of the cereal ß-glucans, the
basis for any observed differences in physical properties, if related to
structure rather than to molecular weight, probably involves differences in fine structure. The presence of sequences of (1→4) linkages
greater than three is now established for oats.

TABLE 2
Molar Ratios of Tri- and Tetrasaccharides Released by Lichenase
from Oats, Barley, and Wheat[a,b]

Cereal	Number of Samples	Molar Ratio	
		Mean ± SD	Range
Oats (Avena sativa)	11[c]	2.1 ± 0.1	1.9–2.2
Barley (Hordeum vulgare)	6[c]	2.8 ± 0.2	2.5–3.2
Wheat (Triticum aestivum)	3[d]	3.1 ± 0.4	2.6–3.3

[a]Determined by anion exchange chromatography with pulsed amperometric detection.
[b]Source: Wood et al (1991b) and unpublished data.
[c]Number of different cultivars.
[d]Soft white winter wheat, hard red winter wheat, and hard red spring wheat.

The presence of sequences of (1→3) linkages greater than one is controversial. As discussed above, this structural sequence was not detected by ^{13}C-NMR, but, if present, it should be detected by the technique of Smith degradation, which involves periodate oxidation followed by reduction and mild acid hydrolysis. Because 3-substituted glucopyranose residues are resistant to oxidation by periodate, the technique provides an apparently ideal means to establish the presence of consecutive (1→3) linkages. Without careful identification of products, however, results may be misinterpreted. Consecutive (1→3) linkages were found for oat ß-glucan (Goldstein et al, 1965), but periodate consumption (0.58 mol/mol) was lower than theoretical. (1→3),(1→4)-ß-D-Glucan is oxidized with difficulty (Ishak and Painter, 1971) and, more importantly, even if β-glucan is fully oxidized and reduced, care must be taken to ensure optimum hydrolysis (Dutton and Gibney, 1972; Woodward et al, 1983). According to Woodward et al (1983), the conditions used by Goldstein et al (1965) would give incomplete hydrolysis. The resultant products of higher DP than 2-O-ß-D-glucopyranosyl-D-erythritol might be mistaken for evidence of consecutive (1→3)-linked units. Recently, Vårum and Smidsrød (1988) used hydrolysis conditions close to those recommended by Woodward et al (1983) and did not detect consecutive (1→3) linkages.

Methods other than Smith degradation have been used to detect consecutive (1→3) linkages. Both oat and barley ß-glucans are reported to lose viscosity in the presence of purified endo-(1→3)-ß-D-glucanases (Clarke and Stone, 1966; Moore and Stone, 1972; Bathgate et al, 1974).

The inevitable conclusion remains much the same as Luchsinger's of over 25 years ago (Luchsinger et al 1965)—namely, that results are contradictory and further study is required to establish the presence of consecutive (1→3)-linked units.

Covalent linkage of barley ß-glucan to protein was suggested by Forrest and Wainwright (1977) on the basis of cochromatography of the ß-glucan with amino nitrogen and lability to a protease, thermolysin. Subsequent to these studies, Bamforth et al (1979) suggested that barley contained a "solubilase," tentatively characterized as a carboxypeptidase, that released ß-glucan and disrupted cell walls during malting. Recent studies suggest, however, that fungal cellulase might be the origin of "solubilase" (Yin et al, 1989).

Vårum and Smidsrød (1988) detected charged groups in purified oat ß-glucan by acid-base titration and by osmotic pressure determinations and suggested these might be from protein amino acid residues. Malki and Autio (1990) suggested that viscosity differences between different oat ß-glucans might be due to protein residues and reported a molecular weight loss on trypsin treatment. On the other hand, Forrest and Wainwright (1977) detected no molecular weight change on treatment of barley ß-glucan with trypsin. In our laboratory, treatment of oat ß-glucan with pepsin, trypsin, or chymotrypsin did not alter viscosity (*unpublished data*).

Peptide attachment to cereal ß-glucan might profoundly influence physical properties; the possibility is intriguing, but present evidence is inconclusive. As in past controversies of this nature (Lamport, 1969), characterization of a peptide-polysaccharide linkage would provide indisputable evidence.

Physical Properties

SOLUBILITY AND EXTRACTION CHARACTERISTICS

Oat ß-glucan's value lies in its classification as a viscous gum, or soluble dietary fiber. Since viscosity is a property of liquids, it is the amount of ß-glucan solubilized in a food system that is important, although in food the presence of other hydrated and gelatinized components, and interactions of these, greatly influence rheology. At the fundamental level, both solubility and viscosity are controlled by molecular weight and structure.

Although some polysaccharides, e.g., cellulose, are clearly insoluble in aqueous solvent, there is little evidence to suggest that defined populations of cereal ß-glucans have intrinsically different aqueous solubility characteristics. There are, however, evident differences in extractability.

Prior treatments, such as heating and drying, and method of milling and particle size, as well as extraction variables (solvent, temperature, duration of extraction, liquid-solids ratio) may influence the amount extracted. Endogenous enzymes from untreated oats or barley, whether originating from the cereal itself or from contaminating microorganisms, might increase the amount of ß-glucan extracted (Preece and MacKenzie, 1952; Yin et al, 1989), and prior treatment with hot aqueous ethanol is generally used to avoid this. However, the physical properties of cell walls may be altered by drying (Forrest and Wainwright, 1977; Ballance and Manners, 1978). Commercial heat treatment of oats decreased ß-glucan extractability (Wood et al, 1991c).

Because of differences in procedure, comparison of literature values for solubility, or amounts of ß-glucan extracted, is difficult. Some data on oat and barley ß-glucan are summarized in Table 3. The basis for differences is not clear, but cell wall organization might influence the rate of solubilization. The ß-glucan in the thick subaleurone endosperm cell walls of oats is more slowly extracted than the ß-glucan in the thinner, inner endosperm cell walls (Wood and Fulcher, 1978). However, the endospermic cell walls of wheat, which stain weakly for ß-glucan (Fulcher and Wood, 1983), are thinner than those of oats or barley, and wheat ß-glucan is nevertheless difficult to extract (Beresford and Stone, 1983). It is possible that proportionately more of the wheat ß-glucan is located in the much thicker aleurone cell walls.

The amount of ß-glucan extracted by water at 65°C is often reported for barley, since the amount varies between cultivars and may be related to performance as malt, but few studies indicate whether the extraction was exhaustive. McCleary (1988) reported that ≈90% of the total barley ß-glucan was extracted in successive treatments with water at 40, 65, and 95°C. In contrast, Henry (1985) reported that extraction was 90% complete at 100°C with three successive extractions but that this yielded only 63% of the total barley ß-glucan. Carr et al (1990) found 43 and 44% of ß-glucan extracted by

TABLE 3
Solubilization of Cereal β-Glucan from Oats (O), Oat Bran (OB),
and Barley (B)

		Extraction Conditions				
Study	Solvent	Temp. (ºC)	Number of Extractions × Duration	Liquids-Solids Ratio	Sample	Percent Soluble
Anderson et al (1978)	H2O	65	4 × 30 min	30:1	O	27
					B	24–65
Aastrup (1979)[a]	KCl/HCl, pH 1.5	RT[b]	1 × 1 hr	10:1	B	7–37
Prentice et al (1980)	Na2CO3, pH 10	45	1 × 20 hr	30:1	B	38
Henry (1985)	H2O	100	3 × 10 min	60:1	O	63
					B	66
Åman and Graham (1987)[a]	H2O	38	1 × 2 hr	75:1	O	78
					B	63
Aastrup and Jørgensen (1988)[a]	KCl/HCl, pH 1.5	20	1 × 1 hr	10:1	B	10–36
Shinnick et al (1988)[c]	H2O	100 for 30 min followed by RT for 16 hr		25:1	O	86
					OB	71
Åman et al (1989)[a]	H2O	38	1 × 2 hr	75:1	O	85
					B	65
Carr et al (1990)	H2O	100	1 × 1 hr	50:1	O	43
					OB	44
Wood et al (1991c)	Na2CO3, pH 10	60	1 × 2 hr	200:1	O	72
					OB	73
					B	47

[a]Hot aqueous ethanol extraction not included as a pretreatment of sample.
[b]Room temperature.
[c]Part of dietary fiber analysis scheme; includes termamyl digestion of starch.

water at 100°C from rolled oats and oat bran, respectively. This increased with extent of processing to 57% for "Quick" oats and 75–80% for ready-to-eat products. Changes in cell wall integrity with more vigorous processing were observed microscopically by Yiu et al (1987) and probably explain these results.

Wood et al (1977, 1978) used sodium carbonate at pH 10 and a 10:1 liquid-to-solids ratio to extract oat ß-glucan during three successive 30-min periods at 45°C. Extraction was incomplete. In contrast, Welch and Lloyd (1989) used the same buffer system but a longer duration of extraction (16 hr) and a much higher liquid-solids ratio (200:1) and achieved total extraction from untreated oat flour. A shorter time (2 hr) with the same liquid-solids ratio achieved 70–75% extraction from oats (Wood et al, 1991c). Complete extraction by the same buffer at 80°C over 20 hr with a 30:1 liquid-solids ratio was also reported for both oat and barley ß-glucans (Prentice et al, 1980). In general, alkali extracts more ß-glucan than water, and pH 10 carbonate buffer may give complete extraction under appropriate conditions.

There are, as yet, no satisfactory molecular explanations for apparent differences in ease of extraction. Woodward et al (1988) reported a difference in oligosaccharides released by lichenase from soluble barley ß-glucan at 40 and 65°C, but in that study the presence of disaccharide and glucose in the reaction products makes structural interpretation difficult. Differences in molecular weight may be involved (McCleary, 1988), but cell wall organization rather than molecular properties may be the controlling factor.

For analysis of total ß-glucan, more vigorous extraction treatments have been devised, and these are discussed in a later section.

VISCOSITY

The viscosity of barley ß-glucan reduces barley's value for feed and malt. On the other hand, viscosity is believed to be the basis for oat bran's ability to improve glucose regulation and, possibly, to reduce serum cholesterol. Despite this, the rheology of cereal ß-glucans has not been extensively studied, and viscosities have often been reported without information on shear rate or even concentration. Above low concentrations, the viscosities of solutions of oat ß-glucan are very sensitive to concentration and shear rate. A wide range of shear rates may be encountered in capillary viscometers, which were used in most early determinations of cereal ß-glucan viscosities.

Wood (1986) reported that 1 and 0.73% solutions of oat gum prepared as described by Wood et al (1978) were pseudoplastic (shear sensitive). A typical power law relationship

$$\sigma = K\tau^n$$

was found between sheer stress (σ) and shear rate (τ). The shear rate range (unreported) was 216–1,622 sec⁻¹. Compared to literature values (Elfak et al, 1979), oat gum was more viscous and pseudoplastic than guar gum at 1% (w/v), although the determination of the latter was over a different range of shear rate (14–1,440 sec⁻¹). The power law relationship does not exactly describe the experimental data, and the constants are affected by the range of shear rates chosen, again making comparison of data difficult.

More recently, Wood et al (1990) compared oat gum prepared in a pilot plant (Wood et al, 1989b) with oat gum prepared by a similar extraction process in the laboratory and with commercial guar gum. Each gum contained ≈80% ß-glucan, and solutions were prepared on the basis of polysaccharide concentration. The results (Table 4; shear rate range 1–100 sec⁻¹) showed that, in comparison with commercial guar gum, oat gum prepared in the laboratory was more viscous and pseudoplastic, but material prepared on the kilogram scale in a pilot plant as less. Autio et al (1987), using an oat gum sample prepared essentially as described by Wood et al (1978) and containing similar concentrations of ß-glucan (≈80%), found a similar power law relationship, using a shear rate range of 19–232 sec⁻¹. To obtain comparable results, oat and guar gums were reexamined as before (Wood et al, 1990) but at 1% (w/v) on a solids basis (i.e., 0.8% ß-glucan) and over a shear rate range similar to that used by Autio et al (1987) (Table 4). The viscosity characteristics of the laboratory-prepared oat gum and the oat gum of Autio et al (1987) were similar. Under these conditions of measurement, the power law more closely described the behavior of the laboratory oat gum and guar gum than that of the pilot-plant oat gum.

TABLE 4
Viscosity Characteristics, Based on the Power Law Equation $\sigma = K\tau^n$, of Oat[a] and Guar Gum Solutions, 1% (w/v, dry wt solids basis) in Water

Elfak et al (1979)	Guar	14–1,440	0.28	10
Wood (1986)	Oat (L)	216–1,622	0.19	51
Autio et al (1987)	Oat	19–232	0.28	24
Wood et al (1990)	Oat (PP)[d]	1–100	0.66	3.2
	Oat (L)[d]	1–100	0.27	31
	Guar[d,e]	1–100	0.39	12
Wood (*unpublished data*)	Oat (PP)	18–235	0.59	1.6
	Oat (L)	18–235	0.22	21
	Guar[e]	18–235	0.36	5.6

[a]Oat (L) = oat gum prepared in laboratory, oat (PP) = oat gum prepared in the pilot plant (POS, Saskatoon, SK).
[b]n = flow behavior index, a measure of deviation from Newtonian ($n = 1$) behavior.
[c]K = consistency index, a theoretical value of viscosity (Pa·sec) at a shear rate of 1 sec⁻¹.
[d]Solution was 1% on the basis of polysaccharide concentration.
[e]Guar gum obtained from the Kingsway Chocolate Co., Toronto, Ontario.

The limited range of validity of the power law has been criticized (Morris, 1990), and a more generally valid treatment was proposed for random coil polymers in which two parameters, zero shear viscosity (η_0) and the shear rate $(S_{1/2})$ at which viscosity reduces to one-half zero shear viscosity, completely describe the shear rate behavior. For the pilot-plant oat gum (1%, w/v), a value of 2,000 mPa.sec for η_0 and 32 sec^{-1} for $S_{1/2}$ were found (Wood, *unpublished data*).

Based on low shear rate and viscoelastic measurements, Doublier (1990 and *unpublished*) concluded that the oat gum prepared by Wood et al (1989b) was rheologically similar to guar gum and behaved like a random-coil, non-gelling polymer in aqueous solution at concentrations between 0.1 and 2% (w/v).

MOLECULAR WEIGHT

The previous review (Wood, 1986) noted that little had been published on molecular weight of oat ß-glucan, and this remains true. Vårum and Smidsrød (1988) reported the number average molecular weight (M_n) for some sonicated oat ß-glucan samples and related these to intrinsic viscosity determined in $1M$ LiI. The maximum M_n reported (3-min sonication) was 330,000, and in the equation

$$[\eta] = KM_n^{\alpha} ,$$

values of 5×10^{-4} for K and 0.75 for the exponent were reported.

Size-exclusion chromatography, using a high-performance gel column and Calcofluor postcolumn detection, was recently used to estimate the molecular weight of oat ß-glucans (Wood et al, 1991c). It was shown that use of commercially available pullulans as molecular weight standards led to overestimation of the molecular weight of the oat ß-glucans, and therefore ß-glucans of different molecular weight were used as chromatographic standards. The molecular weights of the chromatographic peaks of these standards were determined by size-exclusion chromatography with low-angle laser light-scattering detection. The results (Table 5) showed a molecular weight for the chromatographic peak of crude extracts from either whole groat or bran of $\approx 3 \times 10^6$, which tended to decrease during isolation and purification. This would give a high value (36 dL/g) for $[\eta]$, using the constants derived by Vårum and Smidsrød (1988). The chromatography, however, was in dilute buffer, whereas the intrinsic viscosity measurements were in $1M$ LiI. Furthermore, the higher molecular weight samples ($>2.2 \times 10^6$) were beyond the range of the standards and additionally subject to error arising from the extreme sensitivity of estimated molecular weight to elution volume.

TABLE 5
Molecular Weights[a] of Partially Purified Oat β-Glucan and Oat β-Glucan
Extracted in pH 10 Carbonate Buffer at 60ºC[b]

Sample	Molecular Weight ($\times 10^{-6}$)
Pilot plant oat gum	1.2
Bench oat gum	2.2
Oat groat extract[c]	2.9
Oat bran extract[c]	3.0

[a]Molecular weight determined from retention time of chromatographic peak using
high-performance size-exclusion chromatography.
[b]Data from Wood et al (1991c).
[c]Average of four cultivars.

Extraction and Purification

Extraction and purification methods were previously reviewed
(Wood, 1986). No new methods have been reported, but details of a
procedure for kilogram-scale production of oat gum were recently
described (Wood et al, 1989b). Viscosity losses occurred throughout
the extraction and isolation in the pilot plant and were greatest
during centrifugation. This latter effect might be shear related, since
the equipment used in the pilot plant generated high shear rates and,
in laboratory experiments, high-speed homogenization and sonication
were shown to reduce viscosity and molecular weight. No satisfactory
explanation was found for the slow loss of viscosity at pH 10. The
viscosity characteristics of the pilot-plant and laboratory oat gums
are compared in Table 4.

Oat gums prepared by carbonate extraction and alcohol precipitation generally contain about 80% ß-glucan when prepared in the
laboratory. Pilot-plant products of similar purity were obtained,
normally in 9–10% yield, from brans containing 12–15% ß-glucan.
This represents a ß-glucan recovery of about 50%. Further purification was achieved by two precipitations with 20% ammonium sulfate
followed by two further precipitations with 2-propanol (Wood et al,
1991b). The product obtained was 96–98% ß-glucan. It is difficult to
assess the absolute purity of (1→3),(1→4)-ß-D-glucan. Quantitative
acid hydrolysis is difficult, and released glucose may arise from
contaminating starch, which must be separately determined enzymatically. Amylolytic enzymes, however, may be contaminated with
interfering ß-glucanase activity. To assess this requires a material that
would serve as the required standard, namely, starch-free ß-glucan.

Analysis

The presence of starch in cereals, in considerably larger amounts
than ß-glucan, presented formidable analytical problems in the past.

No method existed for analysis of the intact polysaccharide, and analysis therefore depended on hydrolysis to glucose. To distinguish between the different origins of the glucose, enzymatic methods were necessary. The difference between total glucose (by acid hydrolysis) and starch glucose (by enzyme hydrolysis) gave ß-glucan (Fleming et al, 1974; Wood et al, 1977). Because measuring a relatively small difference between two large quantities is subject to error, extraction of ß-glucan with minimum starch contamination was necessary. As a consequence, mild methods were used and, because of incomplete extraction, ß-glucan was often underestimated by this approach. To avoid this, a variety of methods using ß-glucanases were developed.

In one approach (Åman and Hesselman, 1985; Åman and Graham 1987), starch was completely degraded by heat-stable α-amylase and amyloglucosidase. ß-Glucan solubilized during the starch digestion was precipitated, and the combined residue and precipitate were treated with a crude ß-glucanase to achieve quantitative conversion to glucose.

In another approach, crude cellulase enzymes were introduced to convert ß-glucan to glucose. A prior extraction remained necessary, since cellulose might interfere, but so long as the cellulase was free from enzymes able to convert starch to glucose, vigorous conditions, such as hydrazine extraction, could be used (Martin and Bamforth, 1981). Simpler extraction methods using 4% sodium hydroxide (25°C, 24 hr, Palmer and MacKenzie [1986]); and dilute perchloric acid (100°C, 3 min, Ahluwalia and Ellis [1984]) have since been proposed. The latter method, which also describes simultaneous starch determination, might have become widely used but for development of a nonenzymatic method (Jørgensen, 1988). The originally recommended (Ahluwalia and Ellis, 1984) perchloric acid extraction time of 3 min may be inadequate. Jørgensen (1988) suggested this be increased to 10 min and used sulfuric instead of perchloric acid, and Carr et al (1990) found that better yields of ß-glucan were obtained in 4% sodium hydroxide (25°C, 16 hr) than in perchloric acid (95°C, 3 min).

Anderson et al (1978) first used lichenase, $(1\rightarrow3),(1\rightarrow4)$-β-D-glucan-4-glucanohydrolase (Moscatelli et al, 1961; Nevins et al, 1978), for quantitation. The use of a purified, specific enzyme overcame problems with extraction and inadequately specific commercial enzyme sources, but it suffered from a somewhat time-consuming oligosaccharide extraction and acid hydrolysis step. These difficulties were solved by the use of ß-glucosidase to convert the oligosaccharides released by lichenase to glucose (McCleary and Glennie-Holmes, 1985). Commercial availability (Megazyme, Aust. Pty. Ltd., North Rocks, Sydney, NSW) of the lichenase and ß-glucosidase in kit form has since established this method as the most widely used and accepted assay for ß-glucan.

Some variations of the lichenase-based assay have been suggested. HPLC has been used to quantitate the tri- and tetrasaccharide reaction products (Wood, 1985; Wood et al, 1991b). This method requires use of a standard ß-glucan or a correction based on the yield of tri- and tetrasaccharides. In another variation, also requiring a ß-glucan standard, oligosaccharides were measured by a reducing sugar assay (Henry, 1984). Since the tri- and tetrasaccharides have equal molar response rather than equal weight response (Henry and Blakeney, 1988), accuracy is dependent on the sample and standard releasing the same proportions of tri-, tetra-, and higher oligosaccharides. As discussed in an earlier section, this is not always the case. Thus, for oats, an oat ß-glucan standard should be used.

Nonenzymatic approaches to analysis (Wood and Weisz, 1984; Jørgensen, 1988) rely on the ability of cereal ß-glucans to bind direct, or cellulose-substantive, dyes such as Calcofluor (Wood and Fulcher, 1978; Wood, 1980). Under appropriate conditions (Wood, 1982), oat ß-glucan is quantitatively and specifically precipitated by Calcofluor, allowing assay as glucose after hydrolysis or colorimetrically (Wood and Weisz, 1984; Welch and Lloyd, 1989). Under conditions in which precipitation does not occur, dye binding manifests as a red shift in absorbance maxima and an increase in absorbance or fluorescence intensity of the dye. This allows sensitive detection of ß-glucan and potential analysis (Wood, 1982). Calcofluor is extremely photosensitive, preventing manual assay, but this problem was overcome by a rapid-flow injection analysis (FIA) system (Jørgensen, 1988).

In a collaborative trial, both the McCleary and Jørgensen assays were acceptable as approved methods (Munck et al, 1989). The enzymatic method gave values 7% lower than the FIA method for typical (4% ß-glucan) barley in the collaborative trial, whereas the regression equation found by Jørgensen and Aastrup (1988) gave identical values. Because insoluble reaction products may form, the McCleary method may underestimate ß-glucan by ≈5%. Previously discussed difficulties in determining the purity of ß-glucan standards are a potential source of error in the FIA assay.

A recent study (Anderson, 1990) described an assay based on binding of Congo red to ß-glucan, but this system has some disadvantages. Congo red may cause precipitation of ß-glucan at low concentrations, and starch interferes (Wood, 1982).

Physiological Effects
of Oat ß-Glucan

Animal and clinical nutritional studies using oatmeal or oat bran are described in Chapters 5 and 6, respectively. This section deals

with studies in which purified or partially purified ß-glucan preparations have been used. Verification, or otherwise, of serum cholesterol reduction and related health benefits that might be derived
from consumption of oat products requires positive identification of
the active ingredient(s) and elucidation of the mechanism. It has
been the assumption, based on studies with purified soluble fibers
extending back to the early 1960s (Keys et al, 1961), that the soluble
fiber of oats, mainly ß-glucan, was the active ingredient (Chen et al,
1981). The evidence that this is the case continues to grow, but there
is also good evidence that other minor components such as the tocotrienols related to vitamin E and certain phenolics may have a similar
effect (Chapter 5).

CHICK GROWTH

Barley feed can inhibit growth of young chicks. This effect has
been attributed to the viscosity of the barley ß-glucan in the gastrointestinal tract (Gohl et al, 1978; White et al, 1981, White et al,
1983). The evidence for this has been indirect, based on improvement
in performance following removal of viscosity by ß-glucanase and by
comparison with the effect of added viscous gums such as guar. Until
recently, the only direct study (White et al, 1981) found that purified
barley ß-glucan (Biocon, Lexington, KY) at 1% of a chick diet produced no significant weight gain difference from a corn diet control.
The duration of this study (three days) was probably too short to
observe significant effects on growth, but the viscosity of intestinal
contents was significantly increased. Inclusion of ß-glucanase with
the ß-glucan in the feed returned intestinal viscosity to close to control values. Commercial ß-glucan is expensive, possibly explaining the
limited duration of this nutritional study, and perhaps more
importantly is significantly less viscous relative to native ß-glucan
and is not manufactured for nutritional studies.

Because of indigestible hull, oats are not normally used for poultry feed, but hull is not a problem with naked cultivars. Cave et al
(1990) determined the feed value for young chicks of naked oats, with
and without added oat gum, and of two oat brans, one treated with hot
ethanol to deactivate enzymes and one untreated (Table 6). Despite a
ß-glucan content (4.2%) similar to that of typical barleys, the feed-
gain ratio of the oats was not significantly different from that of the
corn diet control, nor was it improved by addition of ß-glucanase. In
other experiments (Cave et al, 1990), the amount of naked oats used
(600 g per kilogram of diet) caused a significant increase in feed-gain
ratio (1.85 compared to 1.49 for the control; $P < 0.05$). It is possible
that this variability in feed response arises from differences in

endogenous ß-glucanase activity, which may be microbiological in origin. As discussed earlier, enzyme may increase the amount of ß-glucan extracted and therefore the viscosity of the extract, but concurrent depolymerization counters this concentration-related viscosity increase. An improvement in the feed value of barley during storage has been attributed to activity of microorganisms in the seed (Hesselman et al, 1981). In further support of this concept, chicks fed the deactivated oat bran (13.7% ß-glucan) performed significantly less well than those fed untreated bran (12.2% ß-glucan). It is unlikely that this difference arose from the somewhat higher ß-glucan content in the latter bran. The 20% increase in feed-gain ratio in young chicks fed oats with added oat gum directly demonstrated the effect of isolated cereal ß-glucan on the value of the feed. Oat ß-glucan, in addition to decreasing metabolizable energy by a simple dilution effect, interfered with utilization of nutrients, as shown by significant decreases in bioavailability of lipids and amino acids (Cave et al, 1990).

Welch et al (1988) reported significantly reduced weight gain in chicks fed oat gum (60% ß-glucan) at 3.4% but not at 2.6% of the diet; however, feed intake was not reported. Cave et al (1990) found a significant reduction in weight gain with the naked oat diet (2.5% ß-glucan in the diet), and gain was further reduced by the addition of oat gum (Table 6).

GASTROINTESTINAL EFFECTS

Bégin et al (1989) compared the effects of oat gum prepared by Wood et al (1989b), other soluble dietary fibers, and cellulose on gastrointestinal transit in the rat. All fibers delayed gastric emptying relative to the fiber-free control. The fibers delayed the appearance of

TABLE 6
Effects[a] of Presence of Oat Gum and β-Glucanase Enzyme on Performance
of Male Broiler Chicks, Eight to 20 Days of Age[b]

	Diet						
	Corn- and Soy	Naked Oats	Naked Oats + Oat Gum	Naked Oats + β- Glucanase	Oat Bran	Deactivated Oat Bran	SEM
Feed intake, g	478 a	435 a	377 b	447 a	429 a	307 c	10.3
Weight gain, g	342 a	295 b	211 c	311 d	230 c	149 e	4.3
Feed-gain ratio	1.42 a	1.49 a	1.79 b	1.44 a	1.87 b	2.38 c	0.049

[a]Means followed by the same letter are not significantly different ($P > 0.05$).
[b]Adapted from Cave et al (1990).

the peak of dry matter content in three sections dissected from the small intestine, demonstrating increased intestinal transit time. The difference between dry matter intake in a meal and dry matter content totals in the stomach and intestine, at intervals following the meal, showed that the fibers significantly delayed intestinal absorption.

Postulated mechanisms for the effects of viscous soluble fibers on gastrointestinal function and glucose regulation include an effect of viscosity on the so-called unstirred layer (Johnson and Gee, 1982). Lund et al (1989) demonstrated that oat gum, prepared by Wood et al (1989b), increased the viscosity and moisture content of rat gut contents and increased the apparent thickness of the unstirred layer of the rat intestine (a measure of resistance to diffusion), behaving similarly to guar gum. Cholesterol absorption was inhibited by the gum. Further studies in human volunteers (Johnson and Lund, 1990) have shown that the major fermentable carbohydrate reaching the colon after consumption of cooked oats is ß-glucan. With uncooked oats, a proportion (one third to one half) of the fermentable carbohydrate may arise from undigested starch. Thus, metabolic changes that may be associated with colonic fermentation, such as reduced serum cholesterol, would be related to ß-glucan content but possibly also to cooking or processing.

ß-Glucan in the gastrointestinal tract of rats fed oat gum in their diets was measured. The maximum ß-glucan concentration was found in the ileal section, where most of the starch had been absorbed. Molecular weight of the ß-glucan extracted from the intestinal contents was decreased relative to the diet (Wood et al, 1991c). The molecular weight of oat ß-glucan in digesta of pigs and chicks fed oat diets also declined relative to that of the raw material (Wood et al, *unpublished data*). The basis for these changes is not known. The viscosity of the intestinal contents of fasting rats that had been maintained for 15 days on a diet containing oat gum was greater than that of rats fed a fiber-free diet, although no ß-glucan was detected, suggesting that secretory changes (Bégin et al, 1989) might contribute to intestinal viscosity, which complicates interpretation of such measurements. A number of other factors make assessment of the physiological role of gastrointestinal viscosity difficult. The viscosity of dispersions of foods in viscous gums may not parallel the viscosities of the gums alone (Vachon et al, 1988). It is not known what proportion of oat ß-glucan is solubilized in the stomach and intestine, but in cooked oats, before digestion, the solubilized ß-glucan may represent less than 10% of the total (Yiu et al, 1987). Finally, solution properties of supernatant solutions from extracts, or of purified polysaccharides, do not accurately reflect properties of a real food matrix, in which hydrated cell wall fragments interact with, and retain microstructural associations with, other food components.

EFFECT ON GLYCEMIC RESPONSE

Vachon et al (1988) and Bégin et al (1989) studied the effect of dietary fibers on glycemia in rats. A method was developed that allowed relatively stress-free blood sampling in rats trained to consume a small meal in a short period of time following a 13-hr fast. Neither fasting blood glucose concentration nor postprandial glucose concentration of rats consuming the fiber diets for 15 days was different from concentrations in rats on the fiber-free control diet, but postprandial insulin levels were significantly reduced with soluble fibers. The effects of cellulose and of oat and guar gums are shown in Figure 4. Reduction in peak (30-min) insulin concentrations by soluble fibers was dose responsive. No statistically significant effect of the insoluble fiber cellulose was observed. The fact that cellulose was similar to the soluble fibers in causing delayed gastric emptying and intestinal transit (Bégin et al, 1989) seems to suggest that delayed gastric emptying and increased intestinal transit time are not in themselves responsible for the effects on insulin concentrations.

The effects of oat gum and bran on blood glucose and insulin con-

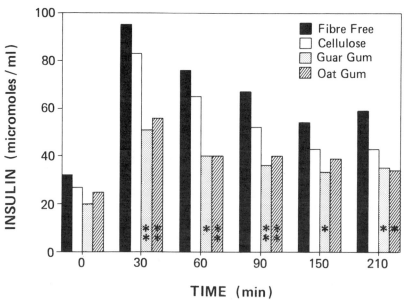

TIME (min)

Figure 4. Effect of oat gum and guar gum at 7.5% of the diet and cellulose at 20% on postprandial insulinemia in the rat. * = $P < 0.05$, ** = $P < 0.01$. (Adapted from Vachon et al, 1988. Reprinted, with permission, from Wood et al, 1989a)

centrations in nondiabetic and type II diabetic subjects have been studied (Wood et al, 1990; Braaten et al, 1991 and *unpublished data*). The effects of oat gum and guar gum were first compared, using the model developed by Jenkins et al (1978), that is, a 50-g glucose load in 500 ml of water containing 14.5 g of gum. This concentration, 2.9% (w/v), forms a very thick, almost gellike mixture. Flavor, color, and aspartame sweetener were added to improve palatability. The postprandial rise in blood glucose and insulin were significantly and similarly reduced by oat and guar gum, compared with the effects of a 50-g glucose load without added gum, as shown in Figure 5.

In more recent studies (Wood et al, *unpublished data*), the postprandial blood glucose and insulin responses to different oat gum doses (7.3, 3.6, and 1.8 g) were examined using the same experimental model. The mean peak plasma glucose increments above baseline, 1.8, 2.1, and 2.5 mmol/ml, respectively, following the 50-g glucose load, even at the lowest dose of 1.8 g of gum, were lower than the increment with glucose alone (3.0 mmol/L). The increment with 7.3 g of gum (1.8 mmol/L) was close to that with 14.5 g (1.9 mmol/L), suggesting that a maximum effective dose had been reached. Mild acid hydrolysis of oat ß-glucan reduced its ability to lower postprandial blood glucose and insulin. An inverse relationship ($r = -0.96$, $P < 0.001$) was found between the incremental peak blood glucose levels and the logarithm of the apparent viscosity of the meal.

In a further experiment, a naturally occurring fiber source, oat bran porridge (OB, 14.6% ß-glucan), was compared with low-fiber Cream of Wheat (CW), in similar hot cereal form. Both nondiabetic and type II diabetic subjects were studied. Each meal was supplemented with white bread to balance available carbohydrate to ≈60 g. The OB meal gave a significantly lower rise in postprandial glucose and insulin levels than the CW meal (Fig. 5) in both nondiabetic and diabetic (not shown) subjects. In a third meal, oat gum was added to the CW to give a ß-glucan dose similar to that of the OB meal. In both subject groups, this meal gave glucose and insulin responses similar to the responses from OB.

REDUCTION IN SERUM CHOLESTEROL

A preparation of oat gum containing 66% ß-glucan was shown to reduce serum cholesterol in hypercholesterolemic rats (Chen et al, 1981) (Table 7). Food intake of the rats decreased 11%, which, although not statistically significant, might have explained the decreased serum cholesterol. The oat gum used in this study was obtained from the Quaker Oat Company (Barrington, IL), which had been assigned a patent (Hyldon and O'Mahoney, 1979) on the use of

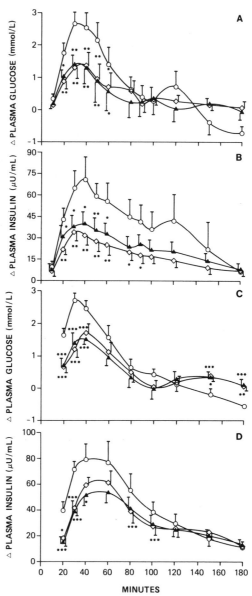

Figure 5. Changes in levels of blood glucose (**A** and **C**) and insulin (**B** and **D**) (difference from baseline, or time 0) in subjects consuming test meals following an overnight fast. **A** and **B,** comparison of oat and guar gums (14.5 g) in 500 ml of water. Circle = glucose alone (control), diamond = glucose + guar gum, triangle = glucose + oat gum. **C** and **D,** comparison of cream of wheat (CW) control with oat bran (OB) and CW + oat gum (OG). Circle = CW, diamond = OB, triangle = CW + OG. Significant difference of values from controls shown as: * = $P < 0.05$, ** = $P < 0.01$. *** = $P < 0.001$. (Reprinted, with permission, from Wood, 1991)

oat gum to treat hypercholesterolemia. Some data from this patent are included in Table 7, since it appears to be the first study of the effect of purified oat gum on serum cholesterol, but lack of information, including statistics, limits the patent's usefulness. The doses of oat gum, calculated from data in the patent by assuming an intake of 18 g of feed per day for rats (weight 265 g), seem very low.

ß-Glucan prepared as described by Wood et al (1989b) reduced the serum cholesterol level in rats when fed at 4% of the diet (Jennings et al, 1988), confirming the earlier results of Chen et al (1981). With the oat gum diet, food intake decreased by 9% relative to the control, but again the difference was not statistically significant. In further unpublished studies (J. W. Anderson, *personal communication*), serum cholesterol was reduced in linear fashion (136, 117, and 101 mg/dl) in rats fed diets containing 3, 6, and 9% gum (2.4, 4.8,

TABLE 7
Effect of Oat Gum on Serum Cholesterol Levels in Rats and Chicks

Animal	Diet	Percent of Diet	β-Glucan in Gum (%)	Serum Cholesterol[a] (mg/dl)	Reference
R	2% Cellulose	0		203	Hyldon and O'Mahoney
		0.15[b]	66[c]	152[d]	(1979)
		0.44[b]	66	140[d]	
		1.03[b]	66	156[d]	
R	10% Cellulose	0		140	Chen et al (1981)
		10[e]	66	83*	
R	White bread[f]	0		104	Klopfenstein and
		7[e]	NR[g]	98	Hoseney (1987)
		13[e]	NR	89*	
R	5% Cellulose	0		151	Jennings et al (1988)
		5	80	117*	
R	15% Wheat bran	0		194	Anderson et al
		3	80	136*	(unpublished data)
		6	80	117*	
		9	80	101*	
C	Low fiber[f]	0		341	Welch et al (1988)
		2.7	60	215*	
	Low fiber[f]	0		297	Welch et al (1988)
		3.4[e]	60	163*	

[a]Asterisks = significantly different from control (0 level); $P < 0.05$.
[b]Calculated from data provided as percent of body weight per day, based on feed intake of 265 g per rat and 18 g per day.
[c]Presumed same as in study of Chen et al (1981).
[d]No statistics reported.
[e]These studies reported reduced feed intake and/or weight gain.
[f]Cholic acid not included in the diet.
[g]Not reported.

and 7.2% ß-glucan), respectively. Food intake (20.3, 18.9, and 19.0 g per day, respectively) was similar for each dose and for the wheat bran control (20 g per day). It seems unlikely, therefore, that reduction in food intake can explain the reduced serum cholesterol concentrations in these experiments.

Klopfenstein and Hoseney (1987) compared serum cholesterol concentrations in rats fed breads incorporating ß-glucan extracts from oats, barley, wheat, and sorghum, but these extracts were not analyzed for ß-glucan. In the described extraction procedure, water-soluble ß-glucan (100°C, 2 hr) was discarded. Furthermore, wheat and sorghum are low in ß-glucan (about 0.8 and 0.1%, respectively). Therefore, some of the cholesterol-lowering effects reported, less than in the other studies in Table 7, were possibly due to pentosan or components other than ß-glucan. This study also differed in using no cholic acid but a higher cholesterol level (5%) in the diets, compared with the other rat studies of Table 7, which used 0.2 and 1%, respectively. Rats fed the oat ß-glucan had lower weight gain and food intake than the control rats.

Serum cholesterol concentrations of young chicks were reduced by oat gum (Welch et al, 1988; Table 7). A protein fraction (containing 1.2% ß-glucan) was also effective but less potent than the gum. No effect from the oil fraction or insoluble residue (starch and fiber) was observed. Oat gum was effective at 3.4 and 2.6% of the diet. Weight gain was significantly less than for the control in the chicks fed 3.4% gum but not in those fed 2.6%.

Conclusions

Oat ß-glucan is composed mainly of ß-(1→3)-linked cellotriosyl and cellotetraosyl units. It is not known how these features are distributed among the polymer chains, although comparison with barley ß-glucan suggests that they are probably randomly arranged. Details of the fine structure of oat ß-glucan reveal features similar to those of barley ß-glucan, such as areas with more than three consecutive (1→4) linkages, but oats have a consistently higher proportion of the cellotetraosyl sequence. New methodology has allowed rapid assessment of this structural feature, which appears to be conserved within *A. sativa*. There remains a need to further investigate possible minor structural features that could profoundly influence physical and functional properties. These might include peptide linkage, consecutive (1→3) linkages, and areas of structural regularity.

Size-exclusion chromatography indicates a molecular weight of 2–3 $\times 10^6$ for extracted oat ß-glucan. Rheological characteristics are similar to those of guar gum, as might be expected for an unbranched, neutral polysaccharide, but more detailed rheological and physical characterization is required.

Problems with analysis of cereal ß-glucan have been largely overcome. Indeed, the sensitivity and specificity of detection and analysis of ß-glucan, based either on enzyme or dye-binding methods, is such as to make oat ß-glucan a useful model in dietary fiber studies.

Like guar gum, oat ß-glucan reduced postprandial glucose and insulin response to an oral glucose challenge in normal human subjects. Oat ß-glucan also moderated the glycemic response when added to a meal of Cream of Wheat and white bread in both normal and diabetic subjects. The effectiveness of oat ß-glucan declined as dose or viscosity was reduced.

The effect of oat ß-glucan on serum cholesterol concentrations in humans is presently under investigation. Four separate studies have shown that oat ß-glucan reduces serum cholesterol in rats, and one study has demonstrated this effect in chicks.

The conclusion is that physiological effects of oat ß-glucan are typical of soluble dietary fiber—namely improvement in glucose regulation and reduction in serum cholesterol. If present studies in humans confirm this, further clinical trials to determine an effective dose of purified ß-glucan and to establish its mechanism of action are warranted.

LITERATURE CITED

AASTRUP, S. 1979. The relationship between the viscosity of an acid flour extract of barley and the ß-glucan content. Carlsberg Res. Commun. 44:289-304.

AASTRUP, S., and JØRGENSEN, K. G. 1988. Application of the Calcofluor flow injection analysis method for determination of ß-glucan in barley, malt, wort, and beer. Am. Soc. Brew. Chem. J. 46:76-81.

AHLUWALIA, B., and ELLIS, E. E. 1984. A rapid and simple method for the determination of starch and ß-glucan in barley and malt. J. Inst. Brew. 90:254-259.

ÅMAN, P., and GRAHAM, H. 1987. Analysis of total and insoluble mixed-linked (1→3)(1→4)-ß-D-glucans in barley and oats. J. Agric. Food Chem. 35:704-709.

ÅMAN, P., and HESSELMAN, K. 1985. An enzymic method for analysis of total mixed-linkage ß-glucans in cereal grains. J. Cereal Sci. 3:231-237.

ÅMAN, P., GRAHAM, H., and TILLY, A.-C. 1989. Content and solubility of mixed-linked (1→3)(1→4)-ß-D-glucan in barley and oats during kernel development and storage. J. Cereal Sci. 10:45-50.

ANDERSON, I. W. 1990. The effect of ß-glucan molecular weight on the sensitivity of dye binding assay procedures for ß-glucan estimation. J. Inst. Brew. 96:323-326.

ANDERSON, M. A., COOK, J. A., and STONE, B. A. 1978. Enzymatic determination of (1→3)(1→4)-ß-glucans in barley grain and other cereals. J. Inst. Brew. 84:233-239.

ASPINALL, G. O., and CARPENTER, R. C. 1984. Structural investigations on the non-starchy polysaccharides of oat bran. Carbohydr. Polym. 4:271-282.

AUTIO, K., MYLLYMAKI, O., and MALKI, Y. 1987. Flow properties of solutions of oat ß-glucans. J. Food Sci. 52:1364-1366.

BALLANCE, G. M., and MANNERS, D. J. 1978. Structural analysis and enzymic solubilization of barley endosperm cell walls. Carbohydr. Res. 61:107-118.

BAMFORTH, C. W., MARTIN, H. L., and WAINWRIGHT, T. 1979. A role for carboxypeptidase in the solubilisation of barley ß-glucan. J. Inst. Brew. 85:334-338.

BATHGATE, G. N., PALMER, G. H., and WILSON, G. 1974. The action of endo-ß-1,3-glucanases on barley and malt ß-glucans. J. Inst. Brew. 80:278-285.

BÉGIN, F., VACHON, C., JONES, J. D., WOOD, P. J., and SAVOIE, L. 1989. Effect of dietary fibers on glycemia and insulinemia and on gastrointestinal function in rats. Can. J. Physiol. Pharmacol. 67:1265-1271.

BERESFORD, G., and STONE, B. A. 1983. (1→3)(1→4)-ß-D-Glucan content of *Triticum* grains. J. Cereal Sci. 1:111-114.

BRAATEN, J. T., WOOD, P. J., SCOTT, F. W., RIEDEL, K. D., POSTE, L. M., and COLLINS, M. W. 1991. Oat gum, a soluble fiber which lowers glucose and insulin in normal individuals after an oral glucose load: Comparison with guar gum. Am. J. Clin. Nutr. 53:1425-1430.

BULIGA, A. S., BRANT, D. A., and FINCHER, G. B. 1986. The sequence statistics and solution conformation of a barley (1→3)(1→4)-ß-D-glucan. Carbohydr. Res. 157:139-156.

BURKITT, D. P., and TROWELL, H. C. 1975. Refined Carbohydrate Food and Disease: Some Implications of Dietary Fibre. Academic Press, London.

CARR, J. M., GLATTER, S., JERACI, J. L., and LEWIS, B. A. 1990. Enzymic determination of β-glucan in cereal-based food products. Cereal Chem. 67:226-229.

CAVE, N. A., WOOD, P. J., and BURROWS, V. D. 1990. Improvement in the nutritive value of naked oats for broiler chicks by various feed additives. Can. J. Anim. Sci. 70:623-633.

CHEN, W.-J. L., ANDERSON, J. W., and GOULD, M. R. 1981. Effects of oat bran, oat gum and pectin on lipid metabolism of cholesterol-fed rats. Nutr. Rep. Int. 24:1093-1098.

CLARKE, A. E., and STONE, B. A. 1966. Enzymic hydrolysis of barley and other ß-glucans by a ß-(1→4)-glucan hydrolase. Biochem. J. 99:582-588.

DAIS, P., and PERLIN, A. S. 1982. High field, 13C-NMR spectroscopy of ß-D-glucans, amylopectin and glycogen. Carbohydr. Res. 100:103-116.

DOUBLIER, J.-L. 1990. Rheological properties of cereal carbohydrates. Pages 111-145 in: Dough Rheology and Baked Product Texture. H. Faridi and J. H. Faubion, eds. AVI-Van Nostrand Reinhold, New York.

DUTTON, G. G. S., and GIBNEY, K. B. 1972. The Smith degradation: A G.L.C. method to monitor the hydrolytic step. Carbohydr. Res. 25:99-105.

ELFAK, A. M., PASS, G., and PHILLIPS, G. O. 1979. The effect of shear rate on the viscosity of solutions of guar gum and locust gum. J. Sci. Food Agric. 30:439-444.

FLEMING, M., and MANNERS, D. J. 1966. A comparison of the fine structure of lichenin and barley glucan. Biochem. J. 100:4P-5P.

FLEMING, M., MANNERS, D. J., JACKSON, R. M., and COOKE, S. C. 1974. The estimation of ß-glucan in barley. J. Inst. Brew. 80:399-404.

FORREST, I. S., and WAINWRIGHT, T. 1977. The mode of binding of ß-glucans and pentosans in barley endosperm cell walls. J. Inst. Brew. 83:279-286.

FULCHER, R. G., and WOOD, P. J. 1983. Identification of cereal carbohydrates by fluorescence microscopy. Pages 111-147 in: New Frontiers in Food Microstructure. D. B. Bechtel, ed. Am. Assoc. Cereal Chem., St. Paul, MN.

GOHL, S., ALDEN, K., ELWINGER, K., and THOMKE, S. 1978. The influence of ß-glucanase on feeding values of barley for poultry and moisture content of excreta. Br. Poult. Sci. 19:41-47.

GOLDSTEIN, I. J., HAY, G. W., LEWIS, B. A., and SMITH, F. 1965. Controlled degradation of polysaccharides by periodate oxidation, reduction and hydrolysis. Pages 361-370 in: Methods in Carbohydrate Chemistry, vol. 5. R. L. Whistler, ed. Academic Press, New York.

HENRY, R. J. 1984. A simplified enzymic method for the determination of $(1{\to}3)(1{\to}4)$-ß-glucans in barley. J. Inst. Brew. 90:178-180.

HENRY, R. J. 1985. A comparison of the non-starch carbohydrates in cereal grains. J. Sci. Food Agric. 36:1243-1253.

HENRY, R. J., and BLAKENEY, A. B. 1988. Evaluation of a general method for measurement of $(1{\to}3)(1{\to}4)$-ß-glucans. J. Sci. Food Agric. 44:75-87.

HESSELMAN, K., ELWINGER, K., NILSSON, M., and THOMKE, S. 1981. The effect of ß-glucanase supplementation, stage of ripeness and storage treatment of barley in diets fed to broiler chickens. Poult. Sci. 60:2664-2671.

HOHNER, G. A., and HYLDON, R. G. 1977. Oat groat fractionation process. U.S. patent 4,028,468.

HYLDON, R. G., and O'MAHONEY, J. S. 1979. Method of treating hypercholesterolemia. U.S. patent 4,175,124.

ISHAK, M. F., and PAINTER, T. 1971. Formation of inter-residue hemiacetals during the oxidation of polysaccharides by periodate ion. Acta Chem. Scand. 25:3875-3877.

JENKINS, D. J. A., WOLEVER, T. M. S., LEEDS, A. R., GASSULL, M. A., HAISMAN, P., DILAWARI, J., GOFF, D. V., METZ, G. L., and ALBERTI, K. G. M. M. 1978. Dietary fibers, fiber analogues, and glucose tolerance: Importance of viscosity. Br. Med. J. 1:1392-1394.

JENNINGS, C. D., BOLEYN, K., BRIDGES, S. R., WOOD, P. J., and ANDERSON, J. W. 1988. A comparison of the lipid-lowering and intestinal morphological effects of cholestyramine, chitosan, and oat gum in rats. Proc. Soc. Exp. Biol. Med. 189:13-20.

JØRGENSEN, K. G. 1988. Quantification of high molecular weight $(1{\to}3)(1{\to}4)$-ß-D-glucan using Calcofluor complex formation and flow injection analysis. I. Analytical principle and its standardisation. Carlsberg Res. Commun. 53:277-285.

JØRGENSEN, K. G., and AASTRUP, S. 1988. Quantification of high molecular weight $(1{\to}3)(1{\to}4)$-ß-D-glucan using Calcofluor complex formation and flow injection analysis. II. Determination of total ß-glucan content of barley and malt. Carlsberg Res. Commun. 53:287-296.

JOHNSON, I. T., and GEE, J. M. 1982. Influence of viscous incubation media on the resistance to diffusion of the intestinal unstirred water layer in vitro. Pfluegers Archiv. 393:139.

JOHNSON, I. T., and LUND, E. K. 1990. Digestibility of starch in cooked and uncooked rolled oats estimated from breath hydrogen (H_2) production in man. J. Physiol. 422:90P.

KAY, R. M., and TRUSWELL, A. S. 1980. Dietary fiber: Effects on plasma and biliary lipids in man. Pages 153-173 in: Medical Aspects of Dietary Fiber. G. A. Spiller and R. M. Kay, eds. Plenum Medical Book, New York.

KEYS, A., GRANDE, F., and ANDERSON, J. T. 1961. Fiber and pectin in the diet and serum cholesterol concentration in man. Proc. Soc. Exp. Biol. Med. 106:555-558.

KLOPFENSTEIN, C. F., and HOSENEY, R. C. 1987. Cholesterol lowering effect of beta-glucan enriched bread. Nutr. Rep. Int. 36:1091-1098.

LAMPORT, D. T. A. 1969. The isolation and partial characterization of hydroxy-proline-rich glycopeptides obtained by enzymic degradation of primary cell walls. Biochemistry 8:1155-1163.

LOCKHART, H. B., and HURT, H. D. 1986. Nutrition of oats. Pages 297-308 in: Oats: Chemistry and Technology. F. H. Webster, ed. Am. Assoc. Cereal Chem., St. Paul, MN.

LUCHSINGER, W. W., CHEN, S.-C., and RICHARDS, A. W. 1965. Mechanism of action of malt beta-glucanases. 9. The structure of barley ß-glucan and the specificity of A_{11}-endo-beta-glucanase. Arch. Biochem. Biophys. 112:531-536.

LUND, E. K., GEE, J. M., BROWN, J. C., WOOD, P. J., and JOHNSON, I. T. 1989. Effect of oat gum on the physicochemical properties of the gastrointestinal contents and on the uptake of D-galactose and cholesterol by rat small intestine in vitro. Br. J. Nutr. 62:91-101.

MALKI, Y., and AUTIO, K. 1990. Oat beta glucan: Physical and nutritional properties. Abstr. 14, 199th National Meeting. American Chemical Society, Washington, DC.

MARTIN, H. L., and BAMFORTH, C. W. 1981. An enzymic method for the measurement of total and water soluble ß-glucan in barley. J. Inst. Brew. 87:88-91.

McCLEARY, B. V. 1988. Purification of $(1\rightarrow3)(1\rightarrow4)$-ß-D-glucan from barley flour. Pages 511-514 in: Methods in Enzymology, vol. 160. W. A. Wood and S. T. Kellogg, eds. Academic Press, San Diego.

McCLEARY, B. V., and GLENNIE-HOLMES, M. 1985. Enzymic quantification of $(1\rightarrow3)(1\rightarrow4)$-ß-D-glucan in barley and malt. J. Inst. Brew. 91:285-295.

MILLER, S. S., PIETRZAK, L. M., WOOD, P. J., and FULCHER, R. G. 1991. Mixed linkage ß-glucan in non-sativa species of Avena. Cereal Chem. In press.

MOORE, A. E., and STONE, B. A. 1972. A ß-1,3-glucan hydrolase from Nicotiana glutinosa. II. Specificity, action pattern and inhibitor studies. Biochim. Biophys. Acta 258:248-264.

MORRIS, E. R. 1990. Shear thinning of 'random coil' polysaccharides: Characterisation by two parameters from a simple linear plot. Carbohydr. Polym. 13:85-96.

MOSCATELLI, E., HAM, E. A., and RICKES, E. L. 1961. Enzymatic properties of a ß-glucanase from Bacillus subtilis. J. Biol. Chem. 236:2858-2862.

MUNCK, L., JØRGENSEN, F. G., RUUD-HANSEN, J., and HANSEN, K. T. 1989. The EBC methods for determination of high molecular weight ß-glucan in barley, malt, wort and beer. J. Inst. Brew. 95:79-82.

NEVINS, D. J., YAMAMOTO, R., and HUBER, D. J. 1978. Cell wall ß-D-glucans of five grass species. Phytochemistry 17:1503-1505.

PALMER, G. H., and MACKENZIE, C. I. 1986. Levels of alkali soluble ß-D-glucans in cereal grains. J. Inst. Brew. 92:461-462.

PARRISH, F. W., PERLIN, A. S., and REESE, E. T. 1960. Selective enzymolysis of poly-ß-D-glucans, and the structure of the polymers. Can. J. Chem. 38:2094-2104.

PATON, D. 1977. Oat starch. I. Extraction, purification and pasting properties. Staerke 29:149-153.

PEAT, S., WHELAN, W. J., and ROBERTS, J. G. 1957. The structure of lichenin. J. Chem. Soc. 3916-3924.

PERLIN, A. S., and SUZUKI, S. 1962. The structure of lichenin: Selective enzymolysis studies. Can. J. Chem. 40:50-56.

PREECE, I. A., and MACKENZIE, K. G. 1952. Non-starchy polysaccharides of cereal grains. II. Distribution of water-soluble gum-like materials in cereals. J. Inst. Brew. 58:457-464.

PRENTICE, N., BABLER, S., and FABER, S. 1980. Enzymic analysis of beta-D-glucans in cereal grains. Cereal Chem. 57:198-202.

SHINNICK, F. L., LONGACRE, M. J., INK, S. L., and MARLETT, J. A. 1988. Oat fiber: Composition versus physiological function in rats. J. Nutr. 118:144-151.

STAUDTE, R. G., WOODWARD, J. R., FINCHER, G. B., and STONE, B. A. 1983. Water soluble (1→3)(1→4)-ß-D-glucans from barley (*Hordeum vulgare*) endosperm. III. Distribution of cellotriosyl and cellotetraosyl residues. Carbohydr. Polym. 3:299-312.

TVAROSKA, I., OGAWA, K., DESLANDES, Y., and MARCHESSAULT, R. H. 1983. Crystalline conformation and structure of lichenan and barley ß-glucan. Can. J. Chem. 61:1608-1616.

VACHON, C., JONES, J. D., WOOD, P. J., and SAVOIE, L. 1988. Concentration effect of soluble dietary fibers on postprandial glucose and insulin in the rat. Can. J. Physiol. Pharmacol. 66:801-806.

VÅRUM, K. M., and SMIDSRØD, O. 1988. Partial chemical and physical characterisation of (1→3)(1→4)-ß-D-glucans from oat (*Avena sativa* L.) aleurone. Carbohydr. Polym. 9:103-117.

WATERMAN, L., and WATERMAN, G. 1989. Forest and Crag: A History of Hiking, Trail-Blazing and Adventure in the Northeast Mountains. Appalachian Mountain Club, Boston, MA.

WELCH, R. W., and LLOYD, J. D. 1989. Kernel (1→3)(1→4)-ß-D-glucan content of oat genotypes. J. Cereal Sci. 9:35-40.

WELCH, R. W., PETERSON, D. M., and SCHRAMKA, B. 1988. Hypocholesterolemic and gastrointestinal effects of oat bran fractions in chicks. Nutr. Rep. Int. 38:551-561.

WHITE, W. B., BIRD, H. R., SUNDE, M. L., PRENTICE, N., BURGER, W. C., and MARLETT, J. A. 1981. The viscosity interaction of barley ß-glucan with *Trichoderma viride* cellulase in the chick intestine. Poult. Sci. 60:1043-1048.

WHITE, W. B., BIRD, H. R., SUNDE, M. L., MARLETT, J. A., PRENTICE, N. A., and BURGER, W. C. 1983. Viscosity of ß-D-glucan as a factor in the enzymatic improvement of barley for chicks. Poult. Sci. 62:853-862.

WILKIE, K. C. B. 1979. The hemicelluloses of grasses and cereals. Pages 215-264 in: Advances in Carbohydrate Chemistry and Biochemistry, vol. 36. R. S. Tipson and D. Horton, eds. Academic Press, New York.

WOOD, P. J. 1980. Specificity in the interaction of direct dyes with polysaccharides. Carbohydr. Res. 85:271-287.

WOOD, P. J. 1982. Factors affecting precipitation and spectral changes associated with complex formation between dyes and ß-D-glucans. Carbohydr. Res. 102:283-293.

WOOD, P. J. 1985. Dye-polysaccharide interactions—Recent research and applications. Pages 267-278 in: New Approaches to Research on Cereal Carbohydrates. R. D. Hill and L. Munck, eds. Elsevier, Amsterdam.

WOOD, P. J. 1986. Oat ß-glucan: Structure, location, and properties. Pages

121-152 in: Oats: Chemistry and Technology. F. H. Webster, ed. Am. Assoc. Cereal Chem., St. Paul, MN.

WOOD, P. J., 1991. Studien zur Anreicherung von β-glucan in Haferspeisekleie und mögliche ernährungs-physiologische Vorteile. Getreide Mehl Brot 45:327-331.

WOOD, P. J., and FULCHER, R. G. 1978. Interaction of some dyes with cereal ß-glucans. Cereal Chem. 55:952-966.

WOOD, P. J., and WEISZ, J. 1984. Use of Calcofluor in analysis of oat beta-D-glucan. Cereal Chem. 61:73-75.

WOOD, P. J., PATON, D., and SIDDIQUI, I. R. 1977. Determination of ß-glucan in oats and barley. Cereal Chem. 54:524-533.

WOOD, P. J., SIDDIQUI, I. R., and PATON, D. 1978. Extraction of high-viscosity gums from oats. Cereal Chem. 55:1038-1049.

WOOD, P. J., ANDERSON, J. W., BRAATEN, J. T., CAVE, N. A., SCOTT, F. W., and VACHON, C. 1989a. Physiological effects of ß-D-glucan rich fractions from oats. Cereal Foods World 34:878-882.

WOOD, P. J., WEISZ, J., FEDEC, P., and BURROWS, V. D. 1989b. Large-scale preparation and properties of oat fractions enriched in (1→3)(1→4)-ß-D-glucan. Cereal Chem. 66:97-103.

WOOD, P. J., BRAATEN, J. T., SCOTT, F. W., RIEDEL, D., and POSTE, L. M. 1990. Comparisons of viscous properties of oat and guar gum and the effects of these and oat bran on glycemic index. J. Agric. Food Chem. 38:753-757.

WOOD, P. J., WEISZ, J., and FEDEC, P. 1991a. Potential for ß-glucan enrichment in brans derived from oat (Avena sativa L.) cultivars of different (1→3),(1→4)-ß-D-glucan concentrations. Cereal Chem. 68:48-51.

WOOD, P. J., WEISZ, J., and BLACKWELL, B. A. 1991b. Molecular characterization of cereal ß-D-glucans. Structural analysis of oat ß-D-glucan and rapid structural evaluation of ß-D-glucans from different sources by high-performance liquid chromatography of oligosaccharides released by lichenase. Cereal Chem. 68:31-39.

WOOD, P. J., WEISZ, J., and MAHN, W. 1991c. Molecular characterization of cereal ß-glucans. II. Size-exclusion chromatography for comparison of molecular weight. Cereal Chem. 68:530-536.

WOODWARD, J. R., and FINCHER, G. B. 1982. Substrate specificities and kinetic properties of two (1→3)(1→4)-ß-D-glucan endo-hydrolases from germinating barley (Hordeum vulgare). Carbohydr. Res. 106:112-122.

WOODWARD, J. R., and FINCHER, G. B. 1983. Water soluble barley ß-glucans. Brewer's Dig. 58(5):28-32.

WOODWARD, J. R., FINCHER, G. B., and STONE, B. A. 1983. Water soluble (1→3)(1→4)-ß-D-glucans from barley (Hordeum vulgare) endosperm. II. Fine structure. Carbohydr. Polym. 3:207-225.

WOODWARD, J. R., PHILLIPS, D. R., and FINCHER, G. B. 1988. Water-soluble (1→3)(1→4)-ß-D-glucans from barley (Hordeum vulgare) endosperm. IV. Comparison of 40°C and 65°C soluble fractions. Carbohydr. Polym. 8:85-97.

YAMAMOTO, R., and NEVINS, D. J. 1978. Structural studies on the ß-D-glucan of the Avena coleoptile cell-wall. Carbohydr. Res. 67:275-280.

YIN, S. S., MACGREGOR, A. W., and CLEAR, R. M. 1989. Field fungi and ß-glucan solubilase in barley kernels. J. Inst. Brew. 95:195-198.

YIU, S. H., WOOD, P. J., and WEISZ, J. 1987. Effects of cooking on starch and ß-glucan of rolled oats. Cereal Chem. 64:373-379.

Chapter 5

Physiological Responses to Dietary Oats in Animal Models

Fred L. Shinnick
The Quaker Oats Company
617 West Main Street
Barrington, Illinois 60010, USA

Judith A. Marlett
Department of Nutritional Sciences
University of Wisconsin-Madison
Madison, Wisconsin 53706, USA

Introduction

Animal models have been useful for investigating the physiological properties of various dietary fiber sources, including oats. The active components and mechanisms of cholesterol reduction, hypoglycemia, laxation, fermentation, and carcinogenesis all have been studied. This chapter reviews the animal studies that have examined the relationships between oats and oat products and these physiological effects. Effects of materials other than oat bran (including barley, pectin, and guar) are also discussed, as they are relevant to understanding the functions of oat bran. The advantages and disadvantages of animal models in this area of research are also considered briefly.

Hypocholesterolemic Properties and Components

Table 1 summarizes investigations of the cholesterol-lowering effects of oat materials in animals. De Groot and co-workers (1963) first reported that rolled oats would reduce serum cholesterol concen-

TABLE 1
Summary of the Cholesterol (C)-Lowering Effects of Oats in Animal Models

Reference	Species	Concentration and Source[a]	Ratio of C to A[b]	Comments
de Groot et al (1963)	Rats	25% rolled oats	1:0.2	Blood C lowered
Fisher and Griminger (1967)	Chicks	43% rolled oats or oat fractions	25% dried eggs	Blood C lowered by ground oats, dehulled flour, and hulls
Forsythe et al (1978)	Rats	70% oat bran	1:0	Blood C reduction not significant
McNaughton (1978)	Chickens	18% ground oats	0:0	Blood C lowered
Chen and Anderson (1979)	Rats	36.5% oat bran	1:0.2	Blood C lowered
Qureshi et al (1980a, 1980b)	Chicks	75% oats, 74% barley	0:0	Blood C lowered more by oats and barley than rye or wheat
Chen et al (1981)	Rats	36.5% oat bran, 10% oat gum	1:0.2	Blood and liver C lowered
Prentice et al (1982)	Chicks	75% oats, 73–74% barley	0:0	Blood C and liver synthetic enzymes lowered
Rogel and Vohra (1983)	Quail	2.3% oat bran, 11.4 % oat hulls	0.5:0	Blood C change not significant
Wilson et al (1984)	Zucker rats	10% oat bran	0:0	Liver C lowered significantly, plasma C lowered nonsignificantly
Schneeman et al (1984)	Rats	20% oat bran	0:0	Blood C not changed
Kritchevsky et al (1984)	Rats	10% oat bran	0.5:0	Blood C not different
Welch et al (1986)	Chicks	10–40% oat bran	0.42:0	Blood and liver reductions dose-dependent
Klopfenstein and Hoseney (1987)	Rats	7% oat β-glucans	77–96 mg of C per day	Blood C reductions not significant
Jennings et al (1988)	Rats	5% oat gum	1:0.2	Blood and liver C lowered
Welch et al (1988)	Chicks	30, 40% oat bran or fractions	14% dried eggs	Blood C lowered by oat bran and gum and protein fractions
Lopez-Guisa et al (1988)	Rats	5–21% oat hulls	1:0.2	Blood and liver C not lowered
Oda et al (1988)	Rats	55% steamed or extruded oats	1:0.25	Extrusion at higher temperature and pressure increased soluble fiber
Shinnick et al (1988)	Rats	26–39% oat bran and processed high-fiber oat flour	1:0.2	Blood and liver C lowered by all oat products
Kahlon et al (1990)	Hamsters	54% oat bran	0.5:0	Blood C lowered
Ranhotra et al (1990)	Rats	50% oat bran or oat bran concentrate	1:0.2	Oat bran concentrate lowered both normal and elevated blood C

(*continued on next page*)

TABLE 1 (continued)

Reference	Species	Concentration and Source[a]	Ratio of C to A[b]	Comments
Shinnick et al (1990)	Rats	0–41% high-fiber oat flour	1:0.1	Blood and liver C reductions were dose-dependent
Nishina et al (1991)	Rats	30% oat bran vs. 8% pectin	0:0	Pectin lowered blood C but not oat bran

[a]Concentration in diet.
[b]Dietary cholesterol (C) and cholic acid (CA) concentrations in percent, or cholesterol source or level.

trations in both rats and humans. Rats were fed semipurified diets containing 15% hydrogenated fat, 1% cholesterol, and 0.2% cholic acid to elevate serum cholesterol. Substitution of rolled oats for wheat starch in the diet resulted in the largest reductions in serum cholesterol concentrations. Rice, instant wheat cereal, whole wheat bread, and whole milk powder also reduced serum cholesterol concentrations, although not as much as rolled oats did. Reliable comparisons of the effectiveness of the various treatments cannot be made because the macronutrient composition of the diets was not uniform and because the weight gain and food intake were not reported and may not have been equal. In humans fed rolled oats, similar observations by the same investigators supported the hypothesis that at least some of the mechanisms involved in the hypocholesterolemic response were similar in humans and rats (de Groot et al, 1963).

In another early study (Fisher and Griminger, 1967), week-old chicks were fed diets containing whole egg powder to elevate serum cholesterol concentrations. Various ground cereal grains and cereal fractions were substituted for cornstarch in the diets to measure their ability to reduce serum cholesterol concentrations. Ground whole oats, dehulled oat flour, oat hulls, ground wheat, and ground barley all lowered serum cholesterol concentrations. Subsequently, several investigators have demonstrated that oats lower blood cholesterol levels in poultry (McNaughton, 1978; Qureshi et al, 1980a,b; Prentice et al, 1982; Rogel and Vohra, 1983; Welch et al, 1986, 1988; Fadel et al, 1987).

In three studies, Anderson and co-workers (Chen and Anderson, 1979; Chen et al, 1981; Jennings et al, 1988) fed rats various soluble fibers. The results demonstrated that oat bran reduced elevated blood cholesterol concentrations, and they strongly indicated that the soluble fiber in oats was the active component (Table 1). Diets contained 6% cottonseed oil, 1% cholesterol, and 0.2% cholic acid to induce hypercholesterolemia. In the first study (Chen and Anderson, 1979), diets contained 10% fiber from pectin, guar gum, or oat bran. All three fibers reduced serum and liver cholesterol concentrations. Pectin and guar gum were somewhat more effective than oat bran, although the oat bran diet would have contained less soluble fiber since less than 50% of the oat bran fiber is soluble (Chapter 3). In the second study (Chen et al, 1981), oat bran and pectin were tested again along with diets containing oat gum. The oat gum was isolated from a coarse bran fraction of oat groats and contained 66% ß-glucan. Pectin reduced serum and liver cholesterol concentrations most, followed by oat gum and then oat bran. The results prompted the authors to suggest that most of the effect of oat bran was due to the ß-glucan concentrated in the oat gum. The third study from this laboratory supported this hypothesis (Jennings et al, 1988). A diet containing 5% oat gum (80% ß-glucan) effectively reduced plasma cholesterol levels in rats.

Shinnick and co-workers (1988, 1990) demonstrated that a "high fiber oat flour" (an improved bran fraction with higher dietary fiber content), obtained by mechanical fractionation of oat bran, lowered hepatic and blood cholesterol levels in the rat. Ney et al (1988) reported that this same material lowered plasma low-density lipoprotein cholesterol concentrations and elevated high-density lipoprotein cholesterol concentrations in rats. The high-fiber oat fractions used in these studies contained about 34% more total dietary fiber than the oat bran, and the mechanical fractionation preferentially increased the portion of the fiber recovered as ß-glucan by 16–31%.

Differences in experimental design can usually account for results that do not show reductions in serum cholesterol concentrations from oat treatments (Table 1). In general, soluble fiber sources, including oats or oat bran, do not lower normal or basal serum cholesterol concentrations in rats or chicks (Ranhotra et al [1990] is an exception). Blood cholesterol concentrations first must be raised, usually with dietary cholesterol and bile salts, or else little or no lowering by an oat diet is observed (Kritchevsky et al, 1984; Schneeman et al, 1984; Wilson et al, 1984; Klopfenstein and Hoseney, 1987; Nishina et al, 1991). When dietary cholesterol is used to elevate blood cholesterol levels, lipid also accumulates in the liver. Some investigators therefore measure the change in hepatic cholesterol concentration, as well as changes in the blood cholesterol level. Cholic acid is needed for maximal cholesterol absorption in rats (Beher et al, 1969; Story et al, 1974), although it does not elevate blood cholesterol in the absence of dietary cholesterol (Shinnick et al, 1990). Thus, even though Forsythe et al (1978) fed rats substantial amounts of oat bran, no reduction in serum cholesterol was observed because hypercholesterolemia had not been induced by inclusion of bile acids in the diet.

Cholesterol reductions are highly associated with the soluble fiber content of the fiber source and with the amount of soluble fiber included in the diet. Failure to consider the compositional differences between purified fiber sources (like pectin and guar gum) and oat bran, which contains starch, protein, and insoluble fiber as well as soluble fiber (Chapter 3), have resulted in levels of feeding that were ineffective in lowering cholesterol (Rogel and Vohra, 1983; Wilson et al, 1984; Nishina et al, 1991). Approximately one third of the dietary fiber of oat bran is soluble, whereas analysis shows all of purified pectin and guar gum to be soluble dietary fiber (although commercial sources are not 100% pure). Oat bran generally contains about 15–20% total dietary fiber (Chapter 3). The diet containing 2.3% oat bran (Rogel and Vohra, 1983) therefore provided only about 0.3–0.5% total dietary fiber concentration in the diet, not enough to affect blood cholesterol concentration. Finally, oat hulls, which are largely an insoluble fiber source and have little or no effect on cholesterol concentrations (Rogel and Vohra, 1983; Lopez-Guisa et al, 1988), must be distinguished from oat bran.

EFFECTS OF PROCESSING

Processing commercial oat bran with combinations of heat, moisture, and pressure does not appear to alter the cholesterol-reducing properties. Humans typically consume oat bran as a hot cooked cereal or baked in muffins or breads, whereas most animal studies have fed minimally processed materials. Ready-to-eat cereals recently have become a dietary source of oat bran that has been subjected to varying degrees of processing. Shinnick et al (1988) measured the plasma cholesterol responses in hypercholesterolemic rats fed oat bran, a high-fiber oat flour, and four ready-to-eat cereals manufactured from the high-fiber oat flour. The four cereals were produced by extrusion at various combinations of pressure. Rats fed the diets containing oat bran had significantly reduced serum and liver cholesterol concentrations relative to hypercholesterolemic control animals (Table 2). There were no differences in food intake among the groups. The small but statistically significant difference in weight gain between the groups fed the high-fiber oat flour A and cereal IV is unlikely to have affected cholesterol concentrations substantially. Clearly, processing had no measurable detrimental effect. The different processes increased the proportion of dietary fiber that was analyzed as soluble fiber by 19–55%, compared with the level in the high-fiber oat flour used (Table 3). This increase in solubility was without effect on serum cholesterol levels in this physiological model, which may not be sensitive to such changes in analytically measured solubility. The reason for the decline in soluble fiber (to the minimum suggested by the AACC definition, Chapter 1) in high-fiber oat flour B is unknown.

Oda and co-workers (1988) investigated oats processed under two different extrusion conditions. They also found that extrusion increased the proportion of the fiber that was recovered as soluble in an enzymatic-gravimetric analysis, but this increase in solubility also

TABLE 2

Hypocholesterolemic Effects in Rats of Unprocessed
and Processed High-Fiber Oat Flour[a,b]

Diet	Daily Food Intake (%)	Daily Weight Gain (%)	Plasma Cholesterol (mg/dl)	Liver Cholesterol (mg/g)
Control[c,d]	16.1 ± 0.9	5.2 ± 0.8 ab	93.4 ± 14.5 a	4.2 ± 0.6 a
Hypercholesterolemic control[d]	16.1 ± 0.9	5.2 ± 0.7 ab	173.8 ± 37.0 d	66.9 ± 6.4 d
Oat bran	16.4 ± 1.1	5.5 ± 0.8 ab	122.7 ± 28.1 bc[e]	39.2 ± 8.1 c
High-fiber oat flour A	16.2 ± 1.5	4.9 ± 0.7 a	119.9 ± 20.5 bc[e]	42.2 ± 8.4 c
Cereal I[f]	16.0 ± 1.1	5.1 ± 0.7 ab	121.6 ± 14.0 bc	32.9 ± 4.9 b
Cereal II[f]	16.2 ± 1.1	5.4 ± 0.8 ab	109.4 ± 24.8 ab[e]	38.1 ± 7.0 c
Cereal III[f]	16.4 ± 0.9	5.3 ± 0.5 ab	109.4 ± 15.4 ab[e]	41.1 ± 6.2 c
Cereal IV[f]	16.2 ± 1.4	5.6 ± 0.8 b	107.7 ± 12.8 ab[e]	31.5 ± 7.9 b

[a]Adapted from Shinnick et al (1988).
[b]Values are means ± standard deviations. Means in a column followed by different letters differ significantly ($P < 0.05$). $n = 12$, except as noted.
[c]All other diets contained 1.0% cholesterol and 0.2% cholic acid.
[d]Contained 5% cellulose; all other diets contained approximately 6% dietary fiber.
[e]$n = 11$.
[f]Cereals I–IV were produced by different extrusion techniques that involved heating to 300–350°F for 1.5–2.0 min. Cereal I was produced from high-fiber oat flour A and cereals II–IV from a different high-fiber oat flour (oat flour B, see Table III). All cereals were ground to 1 mm or less for incorporation into diets.

TABLE 3

β-Glucan and Dietary Fiber Content of Oat Products[a,b]

Oat Product	Soluble β-Glucan (%TF)[c]	Insoluble β-Glucan (%TF)[c]	Total Soluble Fiber (%TF)[c]	Total Insoluble Fiber (%TF)[c]	Total Dietary Fiber[b] (% original dry weight)
Oatmeal	30	5	41	60	12.1 ± 0.2
Oat bran	29	12	39	61	18.6 ± 0.5
High-fiber oat flour A	37	10	43	57	24.9 ± 1.4
High-fiber oat flour B	28	24	33	68	25.0 ± 0.2
Cereal I[d]	34	8	51	49	16.3 ± 0.3
Cereal II[e]	41	10	51	50	18.0 ± 0.2
Cereal III[e]	32	8	45	55	16.8 ± 0.4
Cereal IV[e]	43	6	49	51	24.6 ± 1.0

[a]Adapted from Shinnick et al (1988). β-Glucan was determined as the difference in the sum of glucose and cellobiose between the Cereflo-treated and untreated samples. Soluble and insoluble fiber fractions were separated by centrifugation after solvent extraction and starch digestion in buffer. Results are the means of duplicate analyses.
[b]The remainder of each fiber source consisted of starch, protein, moisture, and ash.
[c]Percent of total fiber.
[d]Prepared from high-fiber oat flour A.
[e]Prepared from high-fiber oat flour B.

did not significantly affect serum cholesterol concentrations in rats. These two studies (Oda et al, 1988; Shinnick et al 1988) as well as cholesterol-reducing trials with humans consuming ready-to-eat oat bran cereals (Anderson et al, 1990; Keenan et al, 1991) strongly suggest that the cholesterol-reducing properties of oat bran are not modified by extrusion processing of oat fractions.

DOSE RESPONSE TO OAT CONSUMPTION

Welch et al (1986) studied the effects of diets incorporating 0–40% oat bran on the cholesterol level in chicks. Two diet designs were used, one with oat bran substituted equally (by weight) for corn-starch and another in which the diets were equal in nitrogen and fat content. Diets contained 0.42% cholesterol. Significant inverse rela-tionships between plasma and liver cholesterol concentrations and oat consumption were demonstrated using both experimental designs (Fig. 1). Decreased weight gain in the groups fed 40% oat bran was not an explanation for the cholesterol reductions since none of the other effective dietary treatments affected weight gain.

A relationship between the amount of oat bran consumed and

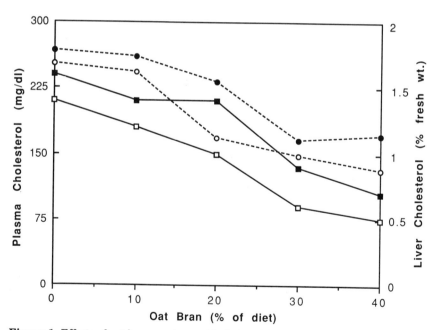

Figure 1. Effects of oat bran on plasma (circles) and liver (squares) cholesterol of rats fed diets containing the indicated amount of oat bran substituted equally for corn-starch (filled symbols) or diets in which the nitrogen and fat were held constant (open symbols) for 14 days. Each symbol represents the mean of eight chicks. (Data from Welch et al, 1986)

decreases in serum and liver lipid concentrations also was studied in hypercholesterolemic rats (Shinnick et al, 1990). Treatment diets contained 0–10%, by weight, dietary fiber from high-fiber oat flour. Highly significant linear inverse relationships were found between the amount of the high-fiber oat flour ingested and the fasting serum and liver cholesterol concentrations (Figs. 2 and 3). Serum triglyceride concentrations also were inversely related to high-fiber oat flour intake ($P = 0.0013$, $r = -0.38$). Weight gain and food intake were not significantly different among any of the treatment or control groups, nor were there significant correlations between fiber intake and weight gain or food intake. The results of both of these studies suggest that even small amounts of oat bran can cause beneficial, albeit small, effects and that larger intakes may result in greater reductions in lipid concentrations.

HYPOCHOLESTEROLEMIC COMPONENTS

Many studies, dating back to Keys (1961), have shown that isolated soluble fibers such as pectin and guar may reduce serum cholesterol. An oat gum preparation was shown to do this in rats (Chen

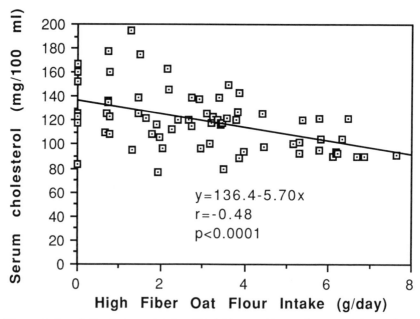

Figure 2. Correlation between intake of high-fiber oat flour and serum cholesterol concentrations in rats. Each point represents the average daily high-fiber oat flour consumption and serum cholesterol concentration of a single rat. $n = 64$. (Reprinted, with permission, from Shinnick et al, 1990; ©American Institute of Nutrition)

et al, 1981). Studies using purified oat ß-glucan preparations, covered in Chapter 4, strongly suggest that this is the most active component in oats. Fisher and Griminger (1967) showed that an isolated oat starch had a small effect on serum cholesterol, suggesting that some component of oat lipids may also contribute to cholesterol lowering. These investigators also demonstrated that isolated oat starch had no effect. Several studies have indicated that plant proteins, including oat protein, are hypocholesterolemic in the diet, at least relative to animal protein sources (Carroll, 1983; Kritchevsky and Czarnecki, 1983; Kritchevsky et al, 1983; Forsythe et al, 1986).

To identify which components were responsible for the choles-terol-lowering effects, Welch et al (1988) separated oat bran into several fractions and demonstrated that the protein and gum fractions were effective hypocholesterolemic agents. The fractionation method was based on that of Wood et al (1977) and generated the following fractions: oil, starch, a protein-enriched fraction, oat gum, and residual solubles (Fig. 4). All of the fractions, including the residual solubles, were dried before incorporation into diets. The gum fraction was 60% ß-glucan and contained about 80% of the ß-glucan from the starting oat bran. None of the other fractions contained more than 3% of the total

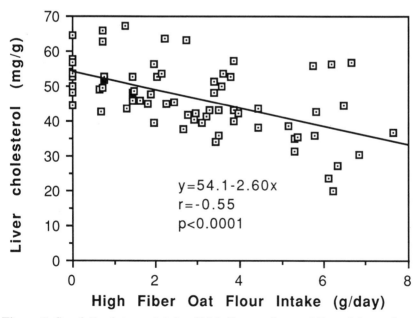

Figure 3. Correlation between intake of high-fiber oat flour and liver cholesterol concentrations in rats. Each point represents the average daily high-fiber oat flour consumption and liver cholesterol concentration of a single rat. n = 64. (Reprinted, with permission, from Shinnick et al 1990; ©American Institute of Nutrition)

ß-glucan. The protein fraction contained nearly 80% protein, which represented about 57% of the protein in the starting material. The insoluble, soluble, and gum fractions contained small portions of the protein of the starting oat bran: 20, 8, and 7%, respectively. Chicks fed a diet containing 14% dried whole eggs as a lipid and cholesterol source were used to test for the cholesterol-lowering ability of the various fractions. Each fraction was fed in proportion to its yield from the starting oat bran, and unfractionated oat bran was fed for comparison. One group was also fed a mixture of all the fractions, recombined in the proportions present in the original bran.

The effects of both 40 and 30% oat bran in the diet were studied. The body weight gain and cholesterol concentrations are summarized in Table 4, expressed as percent of control values for comparison. The insoluble, soluble, and oil fractions had little or no effect on total plasma cholesterol concentration. The gum fraction was most effective, and the protein fraction was intermediate. The mixture of recombined fractions was as effective as the unfractionated oat bran, suggesting that none of the cholesterol-lowering ability had been lost during fractionation. The ß-glucan content (1.2%) of the protein fraction was too low to account for the cholesterol lowering observed. It seems

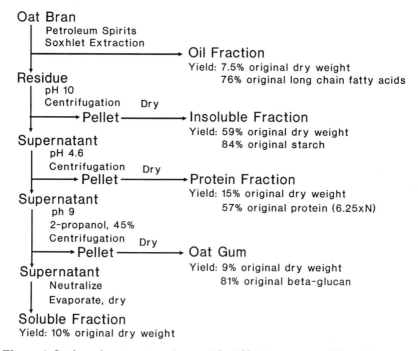

Figure 4. Oat bran fractionation scheme, with yields. (Data from Welch et al, 1988)

likely that the oat protein made some contribution to the effects of oat bran on blood and liver cholesterol concentrations.

Hypocholesterolemic Mechanisms

Various mechanisms have been postulated for the effects of oat bran and other soluble-fiber sources on blood cholesterol concentrations. There are two potential sources of blood and tissue cholesterol: de novo synthesis and dietary absorption; both sources are subject to possible regulatory mechanisms. Regulation of the removal of cholesterol can occur through only one major pathway, conversion to bile acids followed by excretion in feces. Animal studies offer convenient opportunities to test these potential mechanisms.

INHIBITION OF CHOLESTEROL SYNTHESIS

Cholesterol synthesis may be inhibited by the metabolic products of fiber fermentation, short-chain fatty acids (Chen et al, 1984). Soluble fiber sources such as those that lower serum cholesterol concentrations (Pilch, 1987) are rapidly, and usually completely, fermented in the large intestine. Propionate, acetate, and butyrate produced in the cecum and colon are readily absorbed (Schmitt et al, 1976; Cummings, 1981). Anderson and Bridges (1981) demonstrated that propionate inhibited cholesterol synthesis in isolated rat hepatocytes, which suggested a potential mechanism for regulation of blood cholesterol concentrations. Chen et al (1984) demonstrated that dietary propionate (0.5%) reduced serum cholesterol concentrations in rats.

TABLE 4
Effects of Oat Bran or Oat Bran Fractions in Cholesterol-Fed Chicks[a,b]

| Diet Group | 40% Oat Bran[c] | | | 30% Oat Bran[c] | |
	Body Weight	Plasma Cholesterol	Liver Cholesterol	Body Weight	Plasma Cholesterol
Control	100 a	100 a	100 a	100 a	100 a
Native oat bran	100 a	54 c	52 c	89 a	49 c
Remixed components	105 a	54 c	51 c	NT[d]	NT[d]
Insoluble fraction	112 a	93 a	96 a	100 a	102 a
Protein fraction	109 a	70 bc	85 ab	103 a	79 b
Soluble fraction	115 a	84 ab	96 a	101 a	118 a
Gum fraction	75 b	55 c	67 bc	92 a	63 bc
Oil	115 a	84 ab	93 a	98 a	117 a

[a]Data from Welch et al (1988). Values are means of six chicks per diet in the 40% oat bran experiment and seven chicks per diet in the 30% oat bran experiment. Means in the same column followed by different letters are significantly different ($P < 0.05$).
[b]Data are percent of control value.
[c]Concentration of oat bran in diet. Liver cholesterol was not determined in the 30% oat bran experiment.
[d]Not tested.

Storer et al (1983) demonstrated higher concentrations of propionate in hepatic portal venous blood and cecal fluids in rats fed oat bran, as compared to concentrations in cellulose-fed rats, further supporting the existence of a propionate-mediated mechanism.

Illman and Topping (1985) subsequently found, however, that in vivo cholesterol synthesis was increased in rats fed oat bran compared to that in rats fed cellulose-containing diets. This observation is not consistent with a cholesterol-lowering mechanism dependent upon reduced cholesterol synthesis. Further, they observed that hepatic portal venous concentrations of propionate resulting from fermentation were far less than those required for in vitro inhibition of cholesterol synthesis. Additional research by the Australian group supports these observations (Illman et al, 1988). Feeding 5% propionate increased serum propionate concentrations, but never above 1 mM. Cholesterol synthesis was unaffected by 1 mM propionate in perfused liv~rs; 18 mM propionate reduced synthesis, although the rate was still more than half the rate observed without propionate in the perfusate. The authors concluded that propionate has little effect on cholesterol synthesis under normal physiological conditions.

Cholesterol synthesis is also inhibited by organic soluble extracts of nonpolar lipids from plants, which may explain a portion of the cholesterol reduction produced by soluble fiber sources such as oat bran. Researchers at the University of Wisconsin-Madison conducted a series of studies on the effects of barley and oats on lipid metabolism in chickens (Qureshi et al, 1980a; Prentice et al, 1982; Burger et al, 1984). While interpretation of the results from early experiments was complicated by reduced growth on barley and oat diets, it was eventually demonstrated using pair-fed chickens that a high-protein barley could reduce both hydroxymethylglutaryl-coenzyme A (HMG-CoA) reductase activity and serum cholesterol concentrations when growth rates between control and barley groups were equal (Burger et al, 1984). Inclusion in the diet of either a petroleum ether extract of barley or a subsequent methanol extract from the petroleum ether extract caused similar effects on sterol metabolism. They postulated that nonsterol, post-mevalonate metabolites of isoprenoid synthesis in plants could suppress cholesterol synthesis in animals, which would in part explain the cholesterol reductions found with diets high in plant-derived components (Qureshi et al, 1985). One cholesterol synthesis inhibitor identified in the petroleum ether extract of barley was d-α-tocotrienol (Qureshi et al, 1986). Oats also contain significant concentrations of tocotrienols (McLaughlin and Weihrauch, 1979; Cort et al, 1983; Piironen et al, 1986).

Plant sterols, like ß-sitosterol, which are present in oats (Youngs, 1986), may also lower serum cholesterol (Subbiah, 1973). Inhibition

of cholesterol synthesis and reduced intestinal absorption of cholesterol have both been suggested as potential mechanisms for these
plant sterol effects. Although the reported ability of oat oil to lower
cholesterol has been variable (de Groot et al, 1963; Fisher and
Griminger, 1967; Welch et al, 1988), as cited above, various components of the nonpolar fraction may be capable of affecting cholesterol
absorption and metabolism. Further study of these components in both
oats and barley is needed.

In vivo cholesterol synthesis may also be affected by the phos-
phorylation state of HMG-CoA reductase (Kelley and Story, 1987).
Activities of total and expressed HMG-CoA reductase were both reduced
for up to 4 hr after a meal containing oat bran.

Sun et al (1990) found that mechanisms may vary among different fibers. They measured the rate of cholesterol biosynthesis as well
as concentrations of serum and liver total cholesterol in rats after four
weeks of feeding diets containing various fibers plus cholesterol and
cholic acid. Pectin, oat bran, whole amaranth, and defatted amaranth
all reduced serum and liver cholesterol levels. However, rats fed
pectin and whole amaranth incorporated five times as much ^{14}C-acetate
into cholesterol as did rats fed oat bran, suggesting that pectin and
oat bran exert their cholesterol-lowering effects by different mechanisms.

ABSORPTION EFFECTS

Reduced absorption of dietary cholesterol is another postulated
mechanism for the hypocholesterolemic effects of oats. The viscous
nature of oat gum and other soluble fibers suggests that diffusion and
absorption within the gastrointestinal tract may be impaired. Barley
is a cereal grain with physicochemical properties and composition
similar to those of oats. The hypocholesterolemic effect of barley in
chicks is largely lost when diets containing barley are supplemented
with ß-glucanases, which reduce the chain length and viscosity of
ß-glucan (Fadel et al, 1987). Tietyen et al (1990) also reported loss of
cholesterol-lowering potential and viscosity when oat bran treated
with ß-glucanase was fed to rats. Gee et al (1983) demonstrated that
guar gum reduced cholesterol uptake in perfused small intestines of
rats. Similarly, Lund et al (1989) demonstrated that oat gum reduced
cholesterol uptake by isolated everted jejunal sacs. On the other
hand, an investigation with methylcelluloses of varying viscosities
failed to demonstrate any changes in cholesterol synthesis or concentrations of plasma and liver cholesterol (Topping et al, 1988). In
addition, measurements of ^{14}C-cholesterol absorption from single meals
containing either cellulose or oat bran concentrate failed to find any
difference in cholesterol absorption between cellulose, an insoluble fiber,

and oat bran concentrate (Shinnick and Marlett, *unpublished*).

Vahouny et al (1988) have shown that lipid absorption is altered by prolonged fiber feeding, suggesting that some adaptation of the intestinal tract may be involved. Prefeeding guar gum in the diet for 30 days, however, did not change in vitro cholesterol absorption (Gee et al, 1983). Reicks et al (1990) reported that prefeeding rats an oat bran diet also had no effect on the appearance in lymph of cholesterol or long-chain fatty acids, which had been administered intraduodenally in oil.

The effects of ß-glucanase treatment of fiber sources and of tests with isolated tissues are consistent with, but do not prove, that reduced cholesterol absorption may be the mechanism that causes lower serum cholesterol concentrations (Gee et al, 1983; Fadel et al, 1987; Lund et al, 1989; Tietyen et al, 1990). Other investigations, however, have failed to find reductions in cholesterol absorption after feeding soluble fiber (Gee et al, 1983; Reicks et al, 1990). At this time there is no definitive evidence that soluble fiber reduces cholesterol absorption in vivo.

BILE ACID EXCRETION

Another mechanism frequently proposed to explain the cholesterol-lowering effects of dietary fiber sources is increased bile acid excretion arising from increased conversion of cholesterol to bile acids and/or reduced reabsorption of bile acids in the ileum. Reduced bile acid reabsorption may result from binding or trapping of bile acids by fiber or from adaptive changes in intestinal tissue. Illman and Topping (1985) found that dietary oat bran substantially increased the fecal excretion of bile acids and neutral sterols in rats relative to that from cellulose-containing diets. Kritchevsky et al (1984) observed that rats prefed oat bran excreted a greater proportion of a dose of ^{14}C-cholesterol as acidic sterols (bile acids) than the animals prefed cellulose or wheat bran, although the differences were not significant. Increased catabolism of cholesterol and/or excretion of cholesterol metabolites remains a viable mechanism for the hypocholesterolemic properties of dietary fiber.

Other Health Effects

GLYCEMIC EFFECTS

Soluble fiber sources, including oats, reduce postprandial hypoglycemia and insulin secretion (Pilch, 1987; Chapter 4), although the glycemic index (Jenkins et al, 1985) of rolled oats (80%) was much higher than that of foods with a low glycemic index (40%) such as beans. It is possible that viscosity-related changes in physical properties, such as diffusion rates and convective movements in the luminal contents of the digestive tract caused by the soluble fibers, reduce the uptake of monosaccharides. Elsenhans et al (1980) investigated the effects of a series of soluble fibers on monosaccharide and amino acid uptake by rings of everted rat small intestine and found that viscosity was the property that best correlated with the observed reductions in uptake. Ebihara et al (1981) also investigated the effects in rats of soluble fibers that had been shown to reduce early rises in plasma glucose and insulin during glucose tolerance tests in humans. They found that gastric emptying was delayed and the in vitro glucose diffusion rate reduced and suggested that both factors contribute to improved glucose tolerance. Johnson and co-workers have reinforced these observations (Johnson and Gee, 1982; Gee et al, 1983; Lund et al, 1989). They found that dietary oat gum or finely ground rolled oats increased the viscosity of intestinal tract contents and that oat gum reduced galactose uptake by isolated jejunal rings (Lund et al, 1989). The animal evidence is consistent with observations in humans that oats and oat components reduce the rate of glucose uptake, probably in part through increased viscosity of intestinal contents (Pilch, 1987).

FERMENTATION AND LAXATION

There is general agreement that soluble fibers, including those in oats, are highly and rapidly fermentable (Pilch, 1987). Nyman and Asp (1988) investigated the composition and fermentation of the dietary fiber in oat husk (hull), bran, and flour fed to rats. The fiber in the husk, mainly insoluble, was essentially unfermented and explained most of the fecal dry weight increases. The fibers from the bran and flour, on the other hand, were more soluble (38 and 24% of the total dietary fiber, respectively) and fermentable (62 and 55%, respectively). The increase in fecal dry weight that occurred with bran and flour consumption was accounted for less by polysaccharide than by protein and lipid, which is consistent with increased bacterial growth. Fiber digestibility measured in this and many other studies may be underestimated because the polysaccharides of fecal bacteria, which may contain substantial amounts of glucose (Cabotaje et al, 1990), are not subtracted from the fecal analysis. The insoluble fiber content of oats is frequently overlooked but does apparently contribute to fecal bulk (Forsythe et al, 1978).

Differences in production of short-chain fatty acids from fiber fermentation may affect cholesterol synthesis, intestinal cell proliferation, or serum triglycerides. The total amount of short-chain fatty acids produced by fermentation, the proportion of each, and the amount appearing in the blood vary between species. Topping et al (1985) found higher venous short-chain fatty acid concentrations after wheat bran meals than after oat bran meals in pigs, whereas in rats the order was reversed. In addition, butyrate was a much smaller proportion of venous short-chain fatty acids in pigs than in rats.

In an elegant microscopic study using specific histochemical staining techniques, Yiu and Mongeau (1987) studied the breakdown of oat bran in the intestinal tract of the rat. Partially digested oat bran was recovered after salivary digestion and from the ileum and the large intestine. Starch breakdown began quickly in the upper tract. Cell walls rich in ß-glucan largely disappeared from the solid digesta as well, before leaving the upper tract. It was not clear whether this disappearance of ß-glucan was a result of fermentation by the bacteria observed in the sample or simply of solubilization and subsequent removal from the solids, which were recovered for examination. The soluble fraction of digesta was not studied. The aleurone layer remained apparently intact until the material reached the cecum, where its components were rapidly attacked by bacteria. Some of the outer bran tissues, with detectable phenolic content, remained even after cecal fermentation.

A review of the relationship between stool weight and dietary fiber intake (Cummings 1986) indicated that fermentability is only part of the mechanism by which dietary fiber influences laxation. Completely fermented dietary fiber sources had negligible effects on stool weight; dietary fiber sources that were mainly unfermented modestly increased stool weight. The fibers with the greatest effects on stool weight were primarily insoluble but were partially fermented.

CARCINOGENESIS

Epidemiological data suggest that diets high in dietary fiber may have a protective effect against colon cancer (Pilch, 1987). Yet, Jacobs (1990) and Klurfeld (1990), who recently reviewed the effects of dietary fiber on experimental colon cancer in animal models, found that soluble fibers appear to increase intestinal cell proliferation, which has been associated with increased cancer risk. Diets high in soluble fiber tend to cause increases in intestinal length and tissue weight (Pilch, 1987). Oat bran diets increased stomach size and lengthened the small intestine in chicks (Welch et al, 1986), increased the size of the cecum in rats (Lupton et al, 1988), and increased the weight and length of various portions of the intestinal tracts of rats (Schneeman and Richter, 1990).

Jacobs and Lupton have shown, in a series of studies, that diets containing guar gum, pectin, and oat bran (soluble fiber sources) enhance cell proliferation in the small intestine, cecum, and colon in rats (Jacobs, 1983; Jacobs and Lupton, 1984, 1986; Lupton et al,

1988). Oat bran tended to have similar but lesser effects than guar gum and pectin on mucosal cell proliferation. Although the diets contained more oat bran, 20% by weight, than guar gum or pectin, the oat bran diet contained less soluble fiber, about 3%, compared to the 10% fed as guar or pectin.

Pectin, guar gum, and oat bran also substantially increased proximal colonic adenocarcinomas relative to fiber-free diets. Researchers concluded that the type of fiber influenced cell proliferative responses (Jacobs, 1983), that dietary bulk was not a major factor in colonic growth (Jacobs and Lupton, 1984), that tissue damage by dietary fiber did not stimulate cell growth (Lupton et al, 1988), and finally that, in rats, acidification of luminal contents was not protective against dimethylhydrazine-induced colon cancer (Jacobs and Lupton, 1986).

Lupton and Knight (1990) also reported that diets producing lower concentrations of free fatty acids (for instance, diets low in soluble fiber or high in calcium) had lower levels of cell proliferation, suggesting that fatty acids may have a direct role in cell proliferation. Sakata (1987) demonstrated direct effects of short-chain fatty acids on ileal cell proliferation. Johnson and co-workers (Johnson and Gee, 1986; Wyatt et al, 1988) also examined intestinal cell proliferation. They tested less-fermentable soluble fibers and attributed the increased cell proliferation to physical properties of the fibers rather than to fermentation.

The role of soluble fiber sources like oat bran in colon cancer remains unclear (Jacobs, 1990). Whereas a few studies suggest some increased risk of cancer, many other types of evidence suggest either protective effects or no effects (Pilch, 1987). Further, there are substantial differences between chemically induced colon cancer in animals and human colon cancer. Animal colon cancers tend to be proximal rather than distal (Jenkins et al, 1986) and consist of multiple rather than single lesions (Nigro and Bull, 1986).

Summary and Cautions

Various animal species have been used as models in cholesterol and fiber research. Table 5 lists some of the advantages, disadvantages, and references. Animal models of hypercholesterolemia have been a useful adjunct to human experimentation. They will continue to be useful as tools to screen for active materials, evaluate physical or chemical manipulation of active materials, and identify active components. Testing mechanistic theories is another major use of animal models. Experiments can be conducted with animals that are not practical or permitted with humans.

Care must be exercised in the design and execution of any animal

TABLE 5
Relative Utility of Animal Models for Dietary Fiber
and Cholesterol (C) Research

Species	Advantages	Disadvantages	References
Mice	Low cost, numerous inbred strains available	Variable serum C response, limited tissue and fluid, less adaptable to mesh-bottomed cages used for metabolic studies	Paigen et al (1985), Beynen et al (1987), Meijer et al (1987), Kuan and Dupont (1989), Beher et al (1963, 1969)
Gerbils	Low cost, serum C easily elevated	Little genetic variation available, limited previous research, limited tissue and fluids (essentially no urine)	Roscoe and Fahrenbach (1962), Bazzano et al (1971), Mercer and Holub (1981), Nicolosi et al (1981)
Hamsters	Modest cost, serum C easily elevated	Little genetic variation, limited previous research, limited tissue and fluids	Spady et al (1986), Jones and Ridgen (1989), Kahlon et al (1990), Beher et al (1963, 1969), Spady et al (1986), Sakaguchi et al (1987)
Rats	Extensive previous re-search, some genetic variation, very adaptable to research needs, amenable to surgical procedures, modest cost	Serum C may be difficult to elevate, lipoprotein and bile acid metabolism are different from those of other species	See Table I and Beher et al (1963, 1969), Spady et al (1986), Meddings et al (1987), Meijer et al (1987), Sakaguchi et al (1987)
Rabbits	Extensive research use for cardiovascular disease, C easy to ele-vate, more tissue and fluids available	Higher costs	Havel et al (1989), Kritchevsky et al (1989), Osborne et al (1989), Sakaguchi et al (1987)
Pigs	Most like humans (except primates), extensive prior meta-bolic research, ample tissue and fluids available, extensive surgical procedures possible	High costs, animal handling more difficult	Topping et al (1985), Fadel et al (1989)
Chickens	Low cost, extensive prior research, easy to work with	Physiology and ana-tomy very different from humans', sensitive to some diet treatments (i.e., high soluble fiber diets)	See Table I and Fadel et al (1987)

study. Diets must provide adequate and preferably equal amounts of all macro- and micronutrients. Many test materials like oat bran contain substantial amounts of protein, starch, and fat, which can influence experimental results. Induction of hypercholesterolemia in most animal models also requires some manipulation. Balancing that induction to provide cholesterol elevation adequate to allow observation of significant changes in serum cholesterol concentrations without grossly changing the animal's metabolic state requires care. Finally, thought must be given to the time required for changes to take place. Both induction and treatment of hypercholesterolemia appear to be gradual processes involving time-dependent alterations in the metabolic and physiologic states of the animal.

The limitations of animal studies must also be recognized. While, in general, lipid absorption appears to be very similar in most mammals, there are differences in the metabolism of lipids. The distribution and composition of lipoproteins can vary, as can the kinds and proportions of bile acids produced from cholesterol. Researchers must, therefore, be cautious in extending conclusions derived from animal studies to humans.

LITERATURE CITED

ANDERSON, J. W., and CHEN, W.-J. L. 1979. Cholesterol-lowering properties of oat products. Am. J. Clin. Nutr. 32:346-363.

ANDERSON, J. W., SPENCER, D. B., HAMILTON, C. C., SMITH, S. F., TIETYEN, J., BRYANT, C. A., and OELTGEN, P. 1990. Oat-bran cereal lowers serum total and LDL cholesterol in hypercholesterolemic men. Am. J. Clin. Nutr. 52:495-499.

BAZZANO, G. S., WILLIAMS, C., and BAZZANO, G. S. 1971. *In vivo* and *in vitro* inhibition of cholesterol synthesis by glutamic acid feeding. Clin. Res. 19:471.

BEHER, W. T., BAKER, G. D., and PENNEY, D. G. 1963. A comparative study of the effects of bile acids and cholesterol on cholesterol metabolism in the mouse, rat, hamster and guinea pig. J. Nutr. 79:523-530.

BEHER, W. T., FILUS, A. M., RAO, R., and BEHER, M. E. 1969. A comparative study of bile acid metabolism in the rat, mouse, hamster and gerbil. Proc. Soc. Exp. Biol. Med. 130:1067-1074.

BEYNEN, A. C., LEMMENS, A G., DE BRUIJNE, J. J., RONAI, A., WASSMER, B., VON DEIMLING, O., KATAN, M. B., and VAN ZUTPHEN, L. F. M. 1987. Esterases in inbred strains of mice with differential cholesterolemic responses to a high-cholesterol diet. Atherosclerosis 63:239-249.

BURGER, W. C., QURESHI, A. A., DIN, Z. Z., ABUIRMEILEH, N., and ELSON, C. E. 1984. Suppression of cholesterol biosynthesis by constituents of barley kernel. Atherosclerosis 51:75-87.

CABOTAJE, L. M., LOPEZ-GUISA, J. M., SHINNICK, F. L., and MARLETT, J. A. 1990. Neutral sugar composition and gravimetric yield of plant and bacterial fractions of feces. Appl. Environ. Microbiol. 56:1786-1792.

CARROLL, K. K. 1983. Dietary proteins and amino acids—Their effects on cholesterol metabolism. Pages 9-18 in: Animal and Vegetable Proteins in Lipid Metabolism and Atherosclerosis. M. J. Gibney and D. Kritchevsky, eds. Alan R. Liss, New York.

CHEN, W.-J. L., and ANDERSON, J. W. 1979. Effects of plant fiber in decreasing plasma total cholesterol and increasing high-density lipoprotein cholesterol. Proc. Soc. Exp. Biol. Med. 126:108-111.

CHEN, W.-J. L., ANDERSON, J. W., and GOULD, M. R. 1981. Effects of oat bran and pectin on lipid metabolism of cholesterol-fed rats. Nutr. Rep Int. 24:1093-1098.

CHEN, W.-J. L., ANDERSON, J. W., and JENNINGS, D. 1984. Propionate may mediate the hypocholesterolemic effects of certain soluble plant fibers in cholesterol-fed rats. Proc. Soc. Exp. Biol. Med. 175:215-218.

CORT, W. M., VICENTE, T. S., WAYSEK, E. H., and WILLIAMS, B. D. 1983. Vitamin E content of feedstuffs determined by high-performance liquid chromatographic fluorescence. J. Agric. Food Chem. 31:1330-1331.

CUMMINGS, J. H. 1981. Short chain fatty acids in the human colon. Gut 22:763-779.

CUMMINGS, J. H. 1986. The effect of dietary fiber on fecal weight and composition. Pages 211-280 in: CRC Handbook of Dietary Fiber in Human Nutrition. G. A. Spiller, ed. CRC Press, Boca Raton, FL.

DE GROOT, A. P., LUYKEN, R., and PIKAAR, N. A. 1963. Cholesterol lowering effect of rolled oats. Lancet 2:303-304.

EBIHARA, K., MASUHARA, R., and KIRIYAMA, S. 1981. Major determinants of plasma glucose-flattening activity of a water-soluble dietary fiber: Effects of konjac mannan on gastric emptying and intraluminal glucose-diffusion. Nutr. Rep. Int. 23:1145-1156.

ELSENHANS, B., SUJKE, U., BLUME, R., and CASPARY, W. F. 1980. The influence of carbohydrate gelling agents on rat intestinal transport of monosaccharides and neutral amino acids in vitro. Clin. Sci. 59:373-380.

FADEL, J. G., NEWMAN, R. K., NEWMAN, C. W., and BARNES, A. E. 1987. Hypocholesterolemic effects of beta-glucans in different barley diets fed to broiler chicks. Nutr. Rep. Int. 35:1049-1057.

FADEL, J. G., NEWMAN, R. K., NEWMAN, C. W., and GRAHAM, H. 1989. Effects of baking hulless barley on the digestibility of dietary components as measured at the ileum and in the feces in pigs. J. Nutr. 119:722-726.

FISHER, H., and GRIMINGER, P. 1967. Cholesterol-lowering effects of certain grains and of oat fractions in the chick. Proc. Soc. Exp. Biol. Med. 126:108-111.

FORSYTHE, W. A., CHENOWETH, W. L., and BENNINK, M. R. 1978. Laxation and serum cholesterol in rats fed plant fibers. J. Food Sci. 43:1470-1476.

FORSYTHE, W. A., GREEN, M. S., and ANDERSON, J. J. B. 1986. Dietary effects on cholesterol and lipoprotein concentrations: A review. J. Am. Coll. Nutr. 5:533-549.

GEE, J. M., BLACKBURN, N. A., and JOHNSON, I. 1983. The influence of guar gum on intestinal cholesterol transport in the rat. Br. J. Nutr. 50:215-224.

HAVEL, R. J., YAMADA, N., and SHAMES, D. M. 1989. Watanabe heritable hyperlipidemic rabbit. Animal model for familial hypercholesterolemia. Arteriosclerosis 9(Suppl. 1):I33-I38.

ILLMAN, R. J., and TOPPING, D. L. 1985. Effects of dietary oat bran on faecal steroid excretion, plasma volatile fatty acids and lipid synthesis in rats. Nutr. Res. 5:839-846.

ILLMAN, R. J., TOPPING, D. L., McINTOSH, G. H., TRIMBLE, R. P., STORER, G. B., TAYLOR, M. N. and CHENG. B. 1988. Hypocholesterolaemic effects of dietary propionate: Studies in whole animals and perfused rat liver. Ann. Nutr. Metab. 32:97-107.

JACOBS, L. R. 1983. Effects of dietary fiber on mucosal growth and cell proliferation in the small intestine of the rat: A comparison of oat bran, pectin, and guar with total fiber deprivation. Am. J. Clin. Nutr. 37:954-960.

JACOBS, L. R. 1990. Influence of soluble fibers on experimental colon carcinogenesis. Pages 389-401 in: Dietary Fiber: Chemistry, Physiology, and Health Effects. D. Kritchevsky, C. Bonfield, and J. W. Anderson, eds. Plenum Press, New York.

JACOBS, L. R., and LUPTON, J. R. 1984. Effect of dietary fibers on rat large bowel mucosal growth and cell proliferation. Am. J. Physiol. 246:G378-G385.

JACOBS, L. R., and LUPTON, J. R. 1986. Relationship between colonic luminal pH, cell proliferation, and colon carcinogenesis in 1,2-dimethylhydrazine treated rats fed high fiber diets. Cancer Res. 46:1727-1734.

JENKINS, D. J. A., WOLEVER, T. M. S., KALMUSKY, J., GIUDICI, S., GIORDANO, C., WONG, G. S., BIRD, J. N., PATTEN, R., HALL, M., BUCKLEY, G., and LITTLE, J. A. 1985. Low glycemic index carbohydrate foods in the management of hyperlipidemia. Am. J. Clin. Nutr. 42:604-617.

JENKINS, D. J. A., JENKINS, A. L., RAO, A. V., and THOMPSON, L. U. 1986. Cancer risk: Possible protective role of high carbohydrate-high fiber diets. Am. J. Gastroenterol. 81:931-935.

JENNINGS, C. D., BOLEYN, K., BRIDGES, S. R., WOOD, P. J., and ANDERSON, J. W. 1988. A comparison of the lipid-lowering and intestinal morphological effects of cholestyramine, chitosan, and oat gum in rats. Proc. Soc. Exp. Biol. Med. 189:13-20.

JOHNSON, I. T., and GEE, J. M. 1982. Influence of viscous incubation media on the resistance to diffusion of the intestinal unstirred water layer in vitro. Pfluegers Arch. 393:139-143.

JOHNSON, I. T., and GEE, J. M. 1986. Gastrointestinal adaptation in response to soluble non-available polysaccharides in the rat. Br. J. Nutr. 55:497-505.

JONES, P. J. H., and RIDGEN, J. E. 1989. Failure of caloric restriction to influence cholesterol synthesis in hamsters fed identical amounts of dietary cholesterol. Nutr. Res. 9:217-226.

KAHLON, T. S., SAUNDERS, R. M., CHOW, F. I., CHIU, M. M., and BETSCHART, A. A. 1990. Influence of rice bran, oat bran, and wheat bran on cholesterol and triglycerides in hamsters. Cereal Chem. 67:439-443.

KEENAN, J. M., WENZ, J. B., MYERS, S., RIPSIN, C., and HUANG, Z. 1991. Randomized controlled crossover trial of oat bran in hypercholesterolemic subjects. Am. J. Family Practice 33:600-608.

KELLEY, M. J., and STORY, J. A. 1987. Short-term changes in hepatic HMG-CoA reductase in rats fed diets containing cholesterol or oat bran. Lipids 22:1057-1059.

KEYS, A., GRANDE, F., and ANDERSON, J. T. 1961. Fiber and pectin in the diet and serum cholesterol concentration in man. Proc. Soc. Exp. Biol. Med. 106:555-558.

KLOPFENSTEIN, C. F., and HOSENEY, R. C. 1987. Cholesterol-lowering effect of beta-glucan-enriched bread. Nutr. Rep. Int. 36:1091-1098.

KLURFELD, D. M. 1990. Insoluble dietary fiber and experimental colon cancer. Pages 403-415 in: Dietary Fiber: Chemistry, Physiology, and Health Effects. D. Kritchevsky, C. Bonfield, and J. W. Anderson, eds. Plenum Press, New York.

KRITCHEVSKY, D., and CZARNECKI, S. K. 1983. Dietary protein and experimental atherosclerosis: Early history. Pages 1-7 in: Animal and Vegetable Proteins in Lipid Metabolism and Atherosclerosis. M. J. Gibney and D. Kritchevsky, eds. Alan R. Liss, New York.

KRITCHEVSKY, D., TEPPER, S. A., CZARNECKI, S. K., KLURFELD, D. M., and STORY, J. A. 1983. Effects of animal and vegetable protein in experimental atherosclerosis. Pages 85-100 in: Animal and Vegetable Proteins in Lipid Metabolism and Atherosclerosis. M. J. Gibney and D. Kritchevsky, eds. Alan R. Liss, New York.

KRITCHEVSKY, D., TEPPER, S. A., GOODMAN, G. T., WEBBER, M. M., and KLURFELD, D. M. 1984. Influence of oat and wheat bran on cholesterolemia in rats. Nutr. Rep. Int. 29:1353-1259.

KRITCHEVSKY, D., TEPPER, S. A., DAVIDSON, L. M., FISHER, E. A., and KLURFELD, D. M. 1989. Experimental atherosclerosis in rabbits fed cholesterol-free diets. 13. Interaction of proteins and fat. Atherosclerosis 75:123-127.

KUAN, S.-I., and DUPONT, J. 1989. Dietary fat and cholesterol effects on cholesterol metabolism in CBA/J and C57BR/CDJ mice. J. Nutr. 119:349-355.

LOPEZ-GUISA, J. M., HARNED, M. C., DUBIELZIG, R., RAO, S. C., and MARLETT, J. A. 1988. Processed oat hulls as potential dietary fiber sources in rats. J. Nutr. 118:953-962.

LUND, E. K., GEE, J. M., BROWN, J. C., WOOD, P. J., and JOHNSON, I. T. 1989. Effect of oat gum on the physical properties of the gastrointestinal contents and on the uptake of D-galactose and cholesterol by rat small intestine in vitro. Br. J. Nutr. 62:91-101.

LUPTON, J. R., and KNIGHT, D. R. 1990. Effect of calcium, fat and fiber on colonic cell proliferation. FASEB J. 4:A530.

LUPTON, J. R., CODER, D. M., and JACOBS, L. R. 1988. Long-term effects of fermentable fibers on rat colonic pH and epithelial cell cycle. J. Nutr. 118:840-845.

McLAUGHLIN, P. J., and WEIHRAUCH, J. L. 1979. Vitamin E content of foods. J. Am. Diet. Assoc. 75:647-665.

McNAUGHTON, J. L. 1978. Effect of dietary fiber on egg yolk, liver, and plasma cholesterol concentrations of the laying hen. J. Nutr. 108:1842-1848.

MEDDINGS, J. B., SPADY, D. K., and DIETSCHY, J. M. 1987. Kinetic characteristics and mechanisms of regulation of receptor-dependent and receptor-independent LDL transport in the liver of different animal species and humans. Am. Heart J. 113:475-481.

MEIJER, G. W., DE BRUIJNE, J. J., and BEYNEN, A. C. 1987. Dietary cholesterol-fat type combinations and carbohydrate and lipid metabolism in rats and mice. Int. J. Vit. Nutr. Res. 57:319-326.

MERCER, N. J. H., and HOLUB, B. J. 1981. Measurement of hepatic sterol synthesis in the Mongolian gerbil (*Meriones unguiculatus*) *in vivo* using tritium-labeled water: Diurnal variation and effect of type of dietary fat. J. Lipid Res. 22:792-799.

NEY, D. M., LASEKAN, J. B., and SHINNICK, F. L. 1988. Soluble oat fiber tends to normalize lipoprotein composition in cholesterol-fed rats. J. Nutr. 118:1455-1462.

NICOLOSI, R. J., MARLETT, J. A., MORELLO, A. M., FLANAGAN, S. A., and HEGSTED, D. M. 1981. Influence of dietary unsaturated and saturated fat on the plasma lipoproteins of Monogolian gerbils. Atherosclerosis 38:359-371.

NIGRO, N. D., and BULL, A. W. 1986. Dietary studies of cancer of the large bowel in the animal model. Pages 467-479 in: Dietary Fiber: Basic and Clinical Aspects. G. V. Vahouny and D. Kritchevsky, eds. Plenum Press, New York.

NISHINA, P. M., SCHNEEMAN, B. O., and FREEDLAND, R. A. 1991. Effects of dietary fibers on nonfasting plasma lipoprotein and apolipoprotein levels in rats. J. Nutr. 121:431-437.

NYMAN, M. G.-L., and ASP, N.-G. L. 1988. Fermentation of oat fiber in the rat intestinal tract: A study of different cellular areas. Am. J. Clin. Nutr. 48:274-279.

ODA, T., AOE, S., NAKAOKA, M., IDO, K., OHTA, F., and AYANO, Y. 1988. Changes in the dietary fiber content of oats with extrusion cooking and their effect on cholesterol metabolism in rats. J. Jpn. Soc. Nutr. Food Sci. 41:449-456.

OSBORNE, J. A., LENTO, P. H., SIEGFRIED, M. R., STAHL, G. L., FUSMAN, B., and LEFER, A. M. 1989. Cardiovascular effects of acute hypercholesterolemia in rabbits. Reversal with lovastatin treatment. J. Clin. Invest. 83:465-473.

PAIGEN, B., MORROW, A., BRANDON, C., MITCHELL, D., and HOLMES, P. 1985. Variation in susceptibility to atherosclerosis among inbred strains of mice. Atherosclerosis 57:65-73.

PIIRONEN, V., SYVÄOJA, E.-L., VARO, P., SALMINEN, K., and KOIVISTOINEN, P. 1986. Tocopherols and tocotrienols in cereal products from Finland. Cereal Chem. 63:78-81.

PILCH, S. M., ed. 1987. Physiological Effects and Health Consequences of Dietary Fiber. Life Sciences Research Office, Federation of American Societies for Experimental Biology, Bethesda, MD.

PRENTICE, N., QURESHI, A. A., BURGER, W. C., and ELSON, C. E. 1982. Response of hepatic cholesterol, fatty acid synthesis and activities of related enzymes to rolled barley and oats in chickens. Nutr. Rep. Int. 26:597-604.

QURESHI, A. A., BURGER, W. C., PRENTICE, N., BIRD, H. R., and SUNDE, M. L. 1980a. Regulation of lipid metabolism in chicken liver by dietary cereals. J. Nutr. 110:388-393.

QURESHI, A. A., BURGER, W. C., PRENTICE, N., BIRD, H. R., and SUNDE, M. L. 1980b. Suppression of cholesterol and stimulation of fatty acid biosynthesis in chicken livers by dietary cereals supplemented with culture filtrate of *Tricoderma viride*. J. Nutr. 110:1014-1022.

QURESHI, A. A., BURGER, W. C., PETERSON, D. M., and ELSON, C. 1985. Suppression of cholesterogenesis by plant constituents: Review of Wisconsin contributions to NC-167. Lipids 20:817-824.

QURESHI, A. A., BURGER, W. C., PETERSON, D. M., and ELSON, C. E. 1986. The structure of an inhibitor of cholesterol biosynthesis isolated from barley. J. Biol. Chem. 261:10544-10550.

RANHOTRA, G. S., GELROTH, J. A., ASTROTH, K., and RAO, C. S. 1990. Relative lipidemic responses in rats fed oat bran or oat bran concentrate. Cereal Chem. 67:509-511.

REICKS, M., SATCHITHANANDAM, S., and CALVERT, R. J. 1990. Lymphatic fatty acid and cholesterol absorption in wheat or oat bran fed rats. FASEB J. 4:A582.

ROGEL, A. M., and VOHRA, P. 1983. Alteration of lipid metabolism in Japanese quail by feeding oat hulls and brans. Poult. Sci. 62:1045-1053.

ROSCOE, H. G., and FAHRENBACH, M. J. 1962. Cholesterol metabolism in the gerbil. Proc. Soc. Exp. Biol. Med. 110:51-55.

SAKAGUCHI, E., ITOH, H., UCHIDA, S., and HORIGOME, T. 1987. Comparison of fibre digestion and digesta retention time between rabbits, guinea-pigs, rats and hamsters. Br. J. Nutr. 58:149-158.

SAKATA, T. 1987. Stimulatory effect of short-chain fatty acids on epithelial cell proliferation in the rat intestine: A possible explanation for trophic effects of fermentable fibre, gut microbes and luminal trophic factors. Br. J. Nutr. 58:95-103.

SCHMITT, M. G., SOEGREL, K. H., and WOOD, C. M. 1976. Absorption of short chain fatty acids from the human jejunum. Gastroenterology 70:211-215.

SCHNEEMAN, B. O., and RICHTER, B. D. 1990. Long-term feeding of oat bran, wheat bran, or psyllium husk: Effects on the gastrointestinal tract. FASEB J. 4:A528.

SCHNEEMAN, B. P., CIMMARUSTI, J., CHEN, W., DOWNES, L., and LEFEVRE, M. 1984. Composition of high density lipoproteins in rats fed various dietary fibers. J. Nutr. 114:1320-1326.

SHINNICK, F. L., LONGACRE, M. J., INK, S. I., and MARLETT, J. A. 1988. Oat fiber: Composition versus physiological function in rats. J. Nutr. 118:144-151.

SHINNICK, F. L., INK, S. L., and MARLETT, J. A. 1990. Dose response to a dietary oat bran fraction in cholesterol-fed rats. J. Nutr. 120:561-568.

SPADY, D. K., MEDDINGS, J. B., and DIETSCHY, J. M. 1986. Kinetic constants for receptor-dependent and receptor independent low density lipoprotein transport in the tissues of the rat and hamster. J. Clin. Invest. 77:1474-1481.

STORER, G. B., TRIMBLE, R. P., ILLMAN, R. J., SNOSWELL, A. M., and TOPPING, D. L. 1983. Effects of dietary oat bran and diabetes on plasma and caecal volatile fatty acids in the rat. Nutr. Res. 3:519-526.

STORY, J. A., TEPPER, S. A., and KRITCHEVSKY, D. 1974. Influence of synthetic conjugates of cholic acid on cholesterolemia in rats. J. Nutr. 104:1185-1188.

SUBBIAH, M. T. R. 1973. Dietary plant sterols: Current status in human and animal sterol metabolism. Am. J. Clin. Nutr. 26:219-225.

SUN, X.-Q., SHARP, S. W., and LUPTON, J. R. 1990. Cholesterol turnover as a function of fiber type. FASEB J. 4:A527.

TIETYEN, J. L., NEVINS, D. J., and SCHNEEMAN, B. O. 1990. Characterization of the hypocholesterolemic potential of oat bran. FASEB J. 4:A527.

TOPPING, D. L., ILLMAN, R. J., TAYLOR, M. N., and McINTOSH, G. H. 1985. Effects of wheat bran and porridge oats on hepatic portal venous fatty acids in the pig. Ann. Nutr. Metab. 29:325-331.

TOPPING, D. L., OAKENFULL, D., TRIMBLE, R. P., and ILLMAN, R. J. 1988. A viscous fibre (methylcellulose) lowers blood glucose and plasma triacylglycerols and increases liver glycogen independently of volatile fatty acid production in the rat. Br. J. Nutr. 59:21-30.

VAHOUNY, G. V., SATCHITHANANDAM, S., CHEN, I., TEPPER, S. A., KRITCHEVSKY, D., LIGHTFOOT, F. G., and CASSIDY, M. M. 1988. Dietary fiber and intestinal adaptation: Effects on lipid absorption and lymphatic transport in the rat. Am. J. Clin. Nutr. 47:201-206.

WELCH, R. W., PETERSON, D. M., and SCHRAMKA, B. 1986. Hypocholesterolemic, gastrointestinal and associated responses to oat bran in chicks. Nutr. Res. 6:957-966.

WELCH, R. W., PETERSON, D. M., and SCHRAMKA, B. 1988. Hypocholesterolemic and gastrointestinal effects of oat bran fractions in chicks. Nutr. Rep. Int. 38:551-561.

WILSON, J. N., WILSON, S. P., and EATON, R. P. 1984. Dietary fiber and lipoprotein metabolism in the genetically obese Zucker rat. Arteriosclerosis 4:147-153.

WOOD, P. J., PATON, D., and SIDDIQUI, I. R. 1977. Determination of ß-glucan in oats and barley. Cereal Chem. 54:524-533.

WYATT, G. M., HORN, N., GEE, J. M., and JOHNSON, I. T. 1988. Intestinal microflora and gastrointestinal adaptation in the rat in response to nondigestible dietary polysaccharides. Br. J. Nutr. 60:197-207.

YIU, S. H., and MONGEAU, R. 1987. Fluorescence and light microscopic analysis of digested oat bran. Food Microstruct. 6:143-150.

YOUNGS, V. L. 1986. Oat lipids and lipid-related enzymes. Pages 205-226 in: Oats: Chemistry and Technology. F. H. Webster, ed. Am. Assoc. Cereal Chem., St. Paul, MN.

Chapter 6

Hypocholesterolemic Effects of Oat Bran in Humans

James W. Anderson
Metabolic Research Group
VA Medical Center
University of Kentucky College of Medicine
Lexington, Kentucky 40511, USA

Susan R. Bridges
VA Medical Center
Lexington, Kentucky 40511, USA

Introduction

Oat products possess unique hypocholesterolemic properties that may have important implications in humans. As early as 1963, de Groot et al, having observed that rolled oats impressively reduced serum cholesterol in rats, found that feeding 140 g of rolled oats per day to healthy young men lowered serum cholesterol 11% in three weeks. Chen and Anderson (1979) also demonstrated that cholesterol-fed rats that received diets supplemented with 10% fiber from oat bran for three weeks had lower plasma and liver cholesterol levels than rats on control diets. In contrast, levels of plasma high-density lipoprotein (HDL) cholesterol of the oat-fed rats increased (Chen et al, 1981). In a preliminary trial with diabetic men, we (Gould et al, 1980) observed that an oat bran supplement of 100 g/day lowered serum cholesterol 36% and increased HDL cholesterol by 82% compared to a control diet. High HDL cholesterol levels appear to exert an antiatherogenic effect in humans and correlate negatively with ischemic heart disease (Castelli et al, 1977; Kannel, 1983). Therapeutic maneuvers that selectively lower total or low-density lipoprotein (LDL) cholesterol levels and increase HDL cholesterol levels are of considerable importance in management of hypercholesterolemia.

The cholesterol-lowering effects of oat bran and oat products in rats and humans have been recently summarized (Anderson and Chen, 1986; Anderson and Siesel, 1990; Anderson et al, 1990a). In this chapter we compare metabolic ward and ambulatory studies of oat bran feeding in humans. We also review proposed mechanisms for the cholesterol-lowering effects of oat bran and discuss the practical implications of oat bran use for reducing the risk of coronary heart disease.

Metabolic Ward Studies

Five metabolic ward studies have carefully compared the effects of oat bran on serum lipid values of hypercholesterolemic men. In a preliminary study with four subjects, Gould et al (1980) explored the potential role of oat bran in regulating serum cholesterol. Incorporating 100 g/day of oat bran into a high-carbohydrate (70% of energy), low-fat (12% of energy) diet decreased the serum cholesterol of these four subjects by 36%. Although the high-carbohydrate, high-fiber and low-fat (HCF) diet alone lowered total cholesterol by 25–30% compared to a control diet, the HCF-oat bran diet was significantly more effective. Only the HCF-oat bran diet improved the ratio of LDL to HDL cholesterol, lowering serum LDL cholesterol by 58%, increasing serum HDL cholesterol by 82%, and reducing the LDL-HDL ratio by 79% (Table 1). In this early trial, however, the HCF diet was lower in total fat, saturated fat, and cholesterol and higher in starch and plant fiber than the control diet. Thus the specific contribution of the oat fraction was difficult to determine. While this study did not provide conclusive results, it provided incentive for more precise examination of the influence of oat bran.

In 1981, Kirby and co-workers (Kirby et al, 1981) evaluated the effects of feeding 100 g/day of oat bran in eight hypercholesterolemic men. The control and oat bran diets were virtually identical in carbohydrate, protein, fat, and cholesterol content and differed only in the inclusion of 100 g of oat bran in the test diet (Table 2). After 10 days, serum total cholesterol levels remained stable on the control diet but fell 13% on the oat bran diet. The oat bran diet also lowered serum LDL cholesterol levels by 14% but did not change serum HDL cholesterol levels. Fecal excretion of total bile acids was 54% higher on the oat bran than on the control diet, but neutral steroid excretion was slightly lower on the oat bran diet.

Subsequently Anderson et al (1984b) reported this LDL-cholesterol-specific effect using 100 g/day of oat bran as a supplement to metabolically controlled diets in 10 hyperlipidemic men. The subjects received a control diet for seven days followed by an oat bran diet for 21 days. The nutrient contents of control and oat bran diets were equivalent, providing 20% of energy as protein, 43% as carbohydrate, and 37% as fat, and containing approximately 430 mg of cholesterol per day (Table 2). The oat bran diet included 100 g of oat bran per day, providing twofold more total fiber and twofold more soluble fiber than the control diet. Serum cholesterol concentrations decreased 20% within 7–11 days of initiation of the oat bran diet and remained at this level through the 21st day (Fig. 1). The oat bran diet decreased serum LDL cholesterol concentrations by 23% and also increased fecal weight and fecal bile acid excretion.

Recently Anderson et al (1991a) compared effects of oat and wheat bran in a metabolic ward study of 20 hypercholesterolemic men, using diets of similar composition (43% carbohydrate, 16% protein, 41% fat, 450 mg of cholesterol). All patients received a control diet (14 g of fiber) for one week, followed by a diet supplemented with either oat bran (30 g of dietary fiber, including 22 g of soluble) or wheat bran (30 g of dietary fiber, including 5 g of soluble) for three weeks. The wheat bran diet did not significantly affect serum cholesterol, LDL cholesterol, or apolipoprotein B-100 levels, but the oat bran diet lowered serum cholesterol levels by 13%, LDL cholesterol levels

TABLE 1
Metabolic Ward Studies Evaluating Effects of Oat Bran
Products on Serum Lipids

Reference Fiber and Source	Subjects (no.)	Dose (g/day)	Days on Diet	Prestudy Chloresterol Level (mg/dl)	Percent Change,[a,b]		
					TC	LDL	HDL
Gould et al (1980) Oat bran	4	100	10–13	278	-36*	-58*	82*
Kirby et al (1981) Oat bran	8	100	10	>260	-13*	-14*	NC[c]
Anderson et al (1984b) Oat bran	10	100	21	>260	-19*	-23*	-5.6
Anderson et al (1991a) Oat bran	10	110	21	200–320	-13*	-15*	-6
Anderson et al (1990c) Oat bran cereal	12	25	14	210–326	-5.4*	-8.5*	-3.3

[a]TC = total cholesterol, LDL = low-density lipoprotein cholesterol, HDL = high-density lipoprotein cholesterol
[b]Values with asterisks are significantly different from control values ($P < 0.05$).
[c]No change.

by 15%, and apolipoprotein B-100 levels by 15%. Both oat and wheat bran diets slightly lowered HDL cholesterol levels.

A ready-to-eat oat bran cereal was evaluated in the metabolic ward with 12 hypercholesterolemic men (Anderson et al, 1990c). Subjects were randomly assigned to either a corn flake or oat bran cereal diet for two weeks, followed by an additional two weeks on the alternate diet. The corn and oat flake diets resembled a typical

TABLE 2
Average Nutrient Composition of Food Eaten Daily on Control Diets and Oat Bran Diets[a]

	Control Diet[b]	Oat Bran Diet[b]
Energy, Kcal	1,954 ± 84	1,954 ± 98
Protein, g	95 ± 5	93 ± 5
Carbohydrate, g	209 ± 9	211 ± 11
Fat, total, g	82 ± 3.8	82 ± 3.8
Saturated, g	30 ± 3.8	32 ± 3.2
Monounsaturated, g	30 ± 1.3	29 ± 1.5
Polyunsaturated, g	18 ± 2.4	17 ± 2.2
Cholesterol, mg	435 ± 11	424 ± 12
Fiber, total, g	20 ± 1.3	33 ± 1
Soluble, g	7 ± 0.6	13 ± 0.3
Oat bran, g	0	94 ± 2.3

[a]Data from Kirby et al (1981), with recalculated fiber values from Anderson and Bridges (1988).
[b]Mean ± SEM.

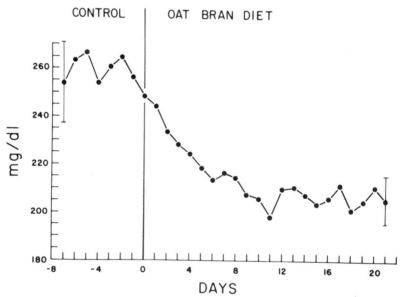

Figure 1. Effect of oat bran on serum cholesterol levels. Six subjects were given 100 g/day of oat bran for 21 days on a metabolic research ward.

American diet (43% carbohydrate, 16% protein, 41% fat, 335 mg of cholesterol), and they were virtually identical in nutrient and energy except for the inclusion of 25 g/day of oat bran in the test diet, providing 8.8 g of total dietary fiber and 3.5 g of soluble fiber. The corn flake cereal provided 0.9 g of total dietary fiber and 0.3 g of soluble fiber. Compared to the corn flake diet, serum total cholesterol decreased 5.4% ($P < 0.05$) with the oat bran cereal diet, LDL cholesterol decreased 8.5% ($P < 0.025$), and serum apolipoprotein B-100 decreased 9.8% ($P < 0.051$). This study again confirmed the hypocholesterolemic properties of oat bran and suggested that these properties are not detrimentally altered by processing.

Ambulatory Studies

Table 3 summarizes eight studies of the effects of oat bran supplements on serum lipids of ambulatory subjects. Further studies were recently reported (Leadbetter et al, 1991).

Storch et al (1984) evaluated the hypocholesterolemic effects of oat bran in an outpatient study with 12 healthy college students. After eating a control diet for two weeks, subjects were randomly assigned to an oat or wheat bran diet for six weeks. Following a seven-week washout period, subjects were crossed over to the other diet. Food intake was assessed weekly by diet diary, 24-hr recall, and seven-day survey. Oat or wheat bran was provided in the form of four muffins daily, which provided a total of 480 kcal and 53 g of bran. Compared to the control diet, oat bran significantly lowered serum cholesterol 12% at three, four, five, and six weeks without reductions in energy or cholesterol and fat intake or changes in activity level.

To evaluate the long-term effects of high-fiber diets rich in water-soluble plant fibers, 10 men with hyperlipidemia incorporated either 100 g of oat bran or 100 g of beans (dry weight) daily into a typical American diet for three weeks while on the metabolic ward. After discharge, the patients used a low-fat, low-cholesterol diet with average intakes of either 41 g of oat bran (dry weight) or 145 g of beans (wet weight) daily. Ten of the men were followed for 24 weeks and four for 99 weeks. At 24 weeks, serum cholesterol was 26% below initial values ($P < 0.001$) and LDL cholesterol was 24% below initial values ($P < 0.001$). As depicted in Figure 2, HDL cholesterol concentrations increased slowly over time up to 99 weeks, whereas LDL cholesterol remained stable over the last 36 weeks, thus decreasing serum LDL-HDL cholesterol ratios (Anderson et al, 1984a).

Van Horn et al (1986) conducted a 12-week outpatient study with 208 healthy adults (aged 30–65 years) to evaluate whether moderate intake of oat bran or oatmeal would enhance the hypocholesterolemic effects of a diet containing low saturated fat and low cholesterol. During the first six weeks, all participants followed the American Heart Association (AHA) fat-modified diet, limiting fat intake to 30% of calories and dietary cholesterol intake to 250 mg/day. At six weeks, subjects were randomly assigned to a diet including oat bran or oatmeal or to a control diet including no oat products for another six-week period. The reported oat product ingestion was about 35–39 g per person per day, with the dietary fat composition remaining similar among the three groups and weight loss being minimal during this period.

Subjects in all three groups experienced a 5.2% reduction in serum cholesterol levels during the first six weeks of the AHA diet alone. During the second six weeks, oat bran or oatmeal supplementation further reduced serum cholesterol by 2.7% and 3.3%, respectively (neither statistically significant), whereas the control group with no oat supplements had less change in serum cholesterol during this time. Differences from baseline values for LDL-cholesterol paralleled those observed for total cholesterol, indicating that including moderate amounts of oat products in the AHA fat-modified diet enhanced serum cholesterol reduction in free-living adults.

TABLE 3
Ambulatory Studies Evaluating Effects of Oat Bran
Products on Serum Lipids

Reference	Subjects (no.)	Dose (g/day)	Days on Diet	Prestudy Chloresterol Level (mg/dl)	Percent Change,[a,b] TC[c]	LDL	HDL
Storch et al (1984)	12	53	42	185	-12*	NR[d]	NR
Anderson et al (1984a)	4	41–100	168	>260	-26*	-24*	-5.3
Van Horn et al (1986)[e]	69	39	42	208	-2.7	NR	NR
Gold and Davidson (1988)	19	34	28	179	-5.3*	-8.7	2.3
Swain et al (1990)	20	87	42	186	0	-2.8	6.1*
Demark-Wahnefried et al (1990)	81	50	84	277	-10.1 to -12.3	NR	2.2 to -5.6
Kestin et al (1990)	24	95	28	245	-4.9*	-6.8*	7.8
Davidson et al (1991)	20	84	42	264	-6.9*	-11.5*	4.3
Keenan et al (1991)	145	56	42	238	-2.2*	-3.9*	0.8

[a]TC = total cholesterol, LDL low-density lipoprotein cholesterol, HDL = high-density lipoprotein cholesterol.
[b]Values with asterisks are significantly different from control values ($P < 0.05$).
[c]Values are expressed as percentage of initial values.
[d]Not reported.
[e]Subjects followed a fat- and cholesterol-restricted diet (American Heart Association) for four to six weeks before receiving oat products.

To evaluate the hypocholesterolemic effects of oat bran in healthy young adults, Gold and Davidson (1988) compared the dietary fiber equivalents of either oat, wheat, or combined wheat and oat brans in a prospective, randomized, double-blind, controlled study. They found that 34 g/day of oat bran for 28 days lowered serum cholesterol 5.3% ($P < 0.05$) and LDL cholesterol 8.7% (nonsignificant), whereas no change was found in control subjects consuming wheat or mixed wheat and oat bran. Oat bran also reduced serum triglyceride levels 8.3%, while levels rose in the other groups.

Anderson et al (1989b) conducted a randomized clinical trial comparing the lipid-lowering effects of an HCF and an AHA Phase II diet. Diet groups differed only in the amount of total fiber intake. The AHA diet provided 20 g and the HCF 25 g of total fiber daily. In this study, 179 healthy men and women (aged 30–50 years) with initial serum cholesterol of 200–300 mg/dl were followed for one year. At the end of the year, all groups, including a nonintervention control group, had sustained significant serum cholesterol reductions. LDL cholesterol dropped significantly in all groups. However, the HCF group, but not the AHA group, had a significantly greater reduction in LDL cholesterol than the control group, suggesting that the HCF diet is more effective than the AHA diet in reducing serum cholesterol, particularly LDL cholesterol.

Swain and colleagues (1990), using a random allocation, cross-

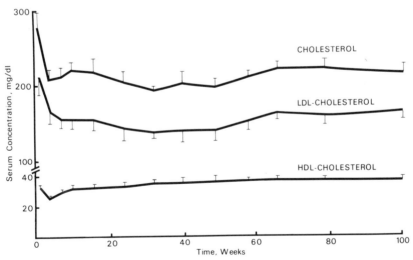

Figure 2. Response of concentrations of serum cholesterol and of low-density lipoprotein (LDL) and high-density lipoprotein (HDL) cholesterol of four hypercholesterolemic men to high-soluble-fiber (one to four weeks) and high-fiber maintenance diets (four to 99 weeks).

over design, compared the effects of oat bran and a low-fiber wheat product fed for six weeks on serum lipids of 20 healthy subjects. Serum total cholesterol, LDL cholesterol, and HDL cholesterol were not significantly different during the two periods. These authors concluded that oat bran has little cholesterol-lowering effect and that high-fiber and low-fiber dietary grain supplements reduce serum cholesterol levels about equally, probably because they replace dietary fats. This study has been criticized (see Anderson and Wood [1990] and other letters in that journal issue) because of these factors: 18 of 20 subjects were unblinded and knew when they were receiving the oat bran; the fat intake during the oat bran phase was 18% higher than during the low-fiber phase; and the statistical analysis was incorrect. Although not reported by the authors, serum HDL cholesterol levels were significantly higher ($P < 0.01$) during the oat-bran phase than during the low-fiber phase.

Demark-Wahnefried and colleagues (1990) examined the effects of oat bran and/or low-fat, low-cholesterol (LFLC) diets on serum lipids of hypercholesterolemic men and women over a 12-week period. Subjects were randomly assigned to one of these four groups: LFLC diet, LFLC diet plus 50 g of oat bran daily; 50 g of oat bran daily without diet modification, or a ready-to-eat oat bran cereal without diet modification. Serum cholesterol values decreased from baseline in all four groups. The greatest reduction occurred with the LFLC diet group (-17.1%), followed by the LFLC diet plus oat bran (-13.1%), oat bran (-12.3%), and oat bran cereal (-10.1%) groups. Table 3 presents changes from baseline for the oat bran and oat bran cereal groups. Because no control group was included and because of decreases in average energy intake and in weight, this study is difficult to interpret.

Kestin and colleagues (1990) rigorously compared the effects of oat bran, rice bran, and wheat bran on serum lipids of 24 mildly hypercholesterolemic men. They used a double-blind crossover design and incorporated the brans into breads and muffins. Only oat bran significantly decreased serum total cholesterol ($P < 0.001$) and LDL cholesterol ($P < 0.001$) compared to those of wheat bran. When changes are compared to baseline values, serum cholesterol decreased 4.9% and LDL cholesterol decreased 6.8% (Table 3).

Davidson and colleagues (1991) carefully examined the effects of oatmeal and oat bran on serum lipid values of 156 hypercholesterolemic adults. Subjects were randomly assigned to the seven following groups: oatmeal, 28, 56, or 84 g per day; oat bran, 28, 56, or 84 g per day; or farina, 28 g per day (control product). Subjects used a LFLC diet for eight weeks before taking the food supplements and continued the diet for six weeks using the oat product or control product. After six weeks, serum cholesterol values had significantly decreased from baseline with these products: oatmeal, 84 g per day (-7.1%, $P = 0.02$); oat bran, 56 g per day, (-9.5%, $P = 0.003$); and oat bran, 84 g per day (-6.9%, $P = 0.03$). Table 3 summarizes lipid responses to 84 g of oat bran daily.

The Keenan study (Keenan et al 1991) was a randomized, controlled, crossover trial of 56 g of oat bran daily for six weeks. Hypercholesterolemic men and women participated, and 145 subjects completed the study. This study showed that the addition of 28 g of oat bran twice daily to an AHA Phase I diet produced a significant reduction in total and LDL cholesterol without significant changes in HDL cholesterol. This study suggested that cholesterol levels in older women were significantly more responsive than levels in women under 50 years old.

Four of these studies—Van Horn et al (1986), Kestin et al (1990), Davidson et al (1991), and Keenan et al (1991)—were very well-controlled studies. The Van Horn study supplied the smallest amount of oat bran daily and failed to demonstrate a significant reduction in serum cholesterol. Whereas all eight studies performed in the United States have used oat bran provided by Quaker Oats, the Kestin study used an Australian-produced oat bran; differences in ß-glucan content between different oat bran preparations could contribute to observed differences in serum cholesterol changes. Nonetheless, these four carefully controlled ambulatory studies indicate that daily intake of oat bran can significantly decrease serum cholesterol values.

Mechanisms

Oat products and other soluble fiber sources may lower serum lipid levels through several mechanisms. Soluble fibers alter cholesterol metabolism at gastrointestinal, hepatic, and peripheral sites (Anderson, 1985). Hypotheses to explain the hypocholesterolemic effects of oat products involve alterations in one or more of the following: cholesterol or bile acid absorption and metabolism, lipid absorption and lipoprotein production, short-chain fatty acid (SCFA) production, and hormone concentrations (Anderson et al, 1990a).

BILE ACID ABSORPTION AND METABOLISM

Alteration in cholesterol and bile acid absorption and reabsorption is likely to be involved in the hypolipidemic mechanism of soluble fiber. A popular hypothesis is that bile acids are sequestered, which is the effect of either fiber binding or viscosity (Story and Kritchevsky, 1976; Vahouny et al, 1980).

Dietary fiber increases bile acid excretion by binding these sterols and preventing their reabsorption. Bile acids are necessary for adequate formation of micelles, through which fat and cholesterol are absorbed. The liver may react to reduced reabsorption by diverting cholesterol from lipoprotein synthesis into bile acid synthesis, thereby secreting less lipoprotein cholesterol and reducing the total body pool of cholesterol (Story and Lord, 1987).

Story and Thomas (1982) suggested that the hypocholesterolemic effect of soluble fiber is related to changes in bile acid pools. A shift in bile acid pools toward chenodeoxycholic acid and its derivatives and away from cholic acid would alter steroid metabolism as a result of the reductions in cholesterol synthesis and absorption induced by chenodeoxycholic acid. In rats, oat bran has been shown to increase fecal excretion of chenodeoxycholic acid and its derivatives and decrease fecal excretion of cholic acid and its derivatives (Story, 1985), thus producing a significant change in concentrations of biliary bile acids.

Plant fiber intake influences biliary bile acid levels in humans as well. Water-soluble fiber (pectin) decreased the ratio of primary to secondary bile acids, whereas insoluble fiber (cellulose) increased primary-to-secondary ratios. Oat bran increased excretion of both cholic and chenodeoxycholic acids in humans (Kirby et al, 1981; Anderson et al, 1984b).

However, the in vitro binding of bile acid by fibers is not well correlated with their hypocholesterolemic effects. Further, not all soluble fibers increase fecal bile excretion, and the magnitude of increase with those that do is small (Kirby et al, 1981; Anderson et al, 1984b). The liver should easily be able to compensate for these small losses. Thus, changes in cholesterol and bile acid absorption cannot fully explain the hypocholesterolemic effects of soluble fiber.

LIPID ABSORPTION AND METABOLISM

Dietary fiber appears to slow lipid digestion, with absorption occurring at more distal sites in the small intestine. Fibers affect sites and rates of absorption of cholesterol, resulting perhaps in gut secretion of lipoprotein particles of different size or composition (Bossello et al, 1984; Redard et al, 1990).

Several physiological effects may interact to decrease lipid and cholesterol digestion in the upper small intestine. As demonstrated in hyperlipidemic individuals, viscous noncellulose polysaccharides appear to reduce plasma cholesterol levels most effectively (Schneeman, 1982), and thus fibers such as guar or oat gum may exert their hypolipidemic effects because of their high viscosity. The so-called unstirred layer adjacent to the mucosa may serve as a physical barrier to nutrient absorption, and impaired physical movement of intestinal contents decreases absorption (Blackburn et al, 1984). The rate of gastric emptying controls the rate of absorption of nutrients from the small intestine. Studies have shown that pectin, which increases viscosity of the gastric contents, delays emptying; cellulose with its negligible effect on viscosity does not affect the emptying rate (Schwartz et al, 1982).

Dietary fiber may also affect digestion and absorption of nutrients within the small intestine through other mechanisms. Certain fibers may alter cell surface and transport mechanisms (Schneeman, 1982; Schneeman and Gallaher, 1985). Altered enzyme activities and pH effects could also decrease digestion of lipids in the upper small intestine (Isaksson et al, 1982).

Water-soluble fibers may alter the rate and site of lipid absorption in the gastrointestinal tract; this in turn could alter the rate of apolipoprotein synthesis by the intestine, a major site of synthesis. Schneeman et al (1984) reported that rats fed oat bran and guar gum showed increases in the percentage of plasma HDL-associated apoprotein A-1 relative to that from wheat bran.

SHORT-CHAIN FATTY ACID CONCENTRATIONS

SCFAs may also mediate the hypocholesterolemic effects of soluble fiber (Chen et al, 1984; Cummings et al, 1987). Soluble fiber is almost completely fermented in the large intestine into SCFAs, including acetic, propionic, and butyric acids (Cummings et al, 1987). SCFAs are found in high concentration in human colonic contents and are present in portal blood (Cummings et al, 1987). Once absorbed, SCFAs pass into the portal vein and to the liver, where propionate and acetate are taken up. While propionate and butyrate are principally metabolized by the colonic mucosa and liver, acetate reaches peripheral tissues, where it is metabolized (Cummings et al, 1987).

Feeding oat bran to rats significantly increases portal vein SCFA concentrations while reducing serum and liver cholesterol levels as compared to levels from feeding cellulose (Chen and Anderson, 1986). Feeding propionate to rats also significantly reduces serum cholesterol (Chen et al, 1984; Illman et al, 1988). While some investigators conclude that propionate does not make a physiological contribution to the hypocholesterolemic effects of oat bran (Illman et al, 1982; Topping et al, 1985), propionate has been reported to inhibit cholesterol synthesis at concentrations of 1.0–2.5 mM in isolated rat hepatocytes (Wright et al, 1990).

Feeding soluble fiber (pectin) elevates peripheral venous acetate in humans (Pomare et al, 1985). Diets supplemented with oat bran sustain a 45% greater increase in serum acetate over a 14-hr period than diets supplemented with wheat bran (Anderson et al, 1989a). Increases in peripheral serum acetate concentrations may inhibit cholesterol synthesis (Beynen et al, 1982) in peripheral tissues, resulting in an increase in peripheral LDL receptors and increased LDL clearance. A higher concentration of peripheral acetate may be accompanied by higher levels of propionate in the portal vein, which could further alter hepatic cholesterol synthesis (Anderson et al, 1989a).

HORMONE CONCENTRATIONS

Carbohydrate and lipid metabolism are closely interrelated. Insulin is a key regulatory hormone for both synthesis and mobilization of lipids. It may promote glucose incorporation into lipids of the arterial wall or foster subtle derangements of serum lipoproteins. Insulin has been reported to increase cholesterol synthesis (Bhathena et al, 1974) and hepatic synthesis and secretion of very-low-density lipoproteins (Reaven and Bernstein, 1978). Therefore, if fiber decreases carbohydrate absorption and insulin secretion, it may indirectly decrease hepatic lipid synthesis.

Soluble fibers have been shown to reduce both glycemic and insulin responses when consumed with meals (Jenkins et al, 1976). Soluble fiber consumed in one meal also appears to improve glucose tolerance to the subsequent meal (Jenkins et al, 1980). Consistent with these studies, Anderson et al (1991b) found that an HCF diet enhanced peripheral insulin sensitivity compared to a low-carbohydrate, low-fiber diet. Improved insulin sensitivity could result, in part, from reduced serum glucagon levels, allowing insulin to act in an unopposed fashion (Munoz et al, 1979).

Soluble fiber, particularly oat fiber, alters pancreatic and gastrointestinal hormone levels, resulting in improved insulin sensitivity and glucose tolerance. These hormonal changes may partially contribute to the hypocholesterolemic effects of certain fibers by reducing synthesis of hepatic cholesterol.

Implications

While not conclusive, epidemiologic evidence suggests that dietary fiber may protect against the development of ischemic heart disease (Anderson et al, 1990b). Khaw and Barrett-Connor (1987) noted that a 6-g increase in daily fiber intake was associated with a 25% reduction in ischemic heart disease mortality. Hypercholesterolemia is the major lipid risk factor for atherosclerosis (Anderson, 1985) and confers an ever-increasing risk for coronary heart disease (CHD)

(Gotto, 1986). Soluble fibers can lower serum cholesterol by 20–30%, theoretically reducing risk for CHD by 40–60% (Anderson et al, 1990b). Diets supplemented with oats and oat bran very effectively lower serum cholesterol and favorably alter the ratio of antiatherogenic (i.e., HDL) to atherogenic (i.e., LDL) lipoproteins in hypercholesterolemic individuals (Gould et al, 1980; Judd and Truswell, 1981; Kirby et al, 1981; Anderson et al, 1984a,b; Turnbull and Leeds, 1987; Van Horn et al, 1988).

As well as reducing lipid risk factors for CHD, dietary fiber indirectly lowers several nonlipid risk factors, including clotting factors, hormonal and glycemic factors, obesity, and hypertension (Anderson and Siesel, 1990). While soluble fiber from foods such as oat products offers several health benefits, these benefits should be considered only within the context of a total healthful dietary plan. Additionally, the lipid-lowering effects of diet are maximized when combined with other positive lifestyle changes, such as increased exercise, smoking cessation, avoidance of alcohol use, and appropriate stress management (Anderson et al, 1990b).

Practical Applications

Hyperlipidemia in most adults relates primarily to diet. Widely accepted diet recommendations to lower serum cholesterol levels include reducing total fat, saturated fat, and cholesterol intake and maintaining a healthy body weight (Connor and Connor, 1977; Bierman and Glomset, 1985). Numerous studies also support the cholesterol-lowering effects of soluble fiber from foods such as oat products (Judd and Truswell, 1981; Turnbull and Leeds, 1987; Van Horn et al, 1988). HCF diets are low in total fat, saturated fat, and cholesterol and high in fiber. They have proven to be a practical and effective approach to the management of hyperlipidemia (Anderson et al, 1984a,b).

Our group developed HCF diets similar to those used in the metabolic ward but suitable for home use for long-term management of diabetes and hyperlipidemia (Anderson and Gustafson, 1988). These diets provide 55–60% of energy as carbohydrate, 15–20% as protein, and 20–25% as fat, as well as less than 200 mg of cholesterol and 50 g of plant fiber daily. They emphasize foods rich in soluble fiber such as oats or bean products (Anderson and Bridges, 1988) as well as other high-fiber foods, including whole-grain cereal products, vegetables, and fruits. They deemphasize red meats, eggs, dairy products, fats, and oils. All of the fiber in such diets originates from commonly available foods; fiber supplements are not used. These diets lower serum cholesterol concentrations by 25–30%, selectively lowering LDL cholesterol 22–25% while maintaining or slightly increasing

HDL cholesterol values (Anderson et al, 1984a,b). Many individuals with hyperlipidemia can avoid drug therapy and normalize their serum cholesterol values by following an HCF diet.

Therapeutic nutrition programs need to be evaluated for efficiency, practicality, palatability, and satiety. We have used HCF diets for 15 years, initially for the management of diabetes and for the past 10 years in the treatment of hyperlipidemia, and we find that they satisfy these criteria. An HCF diet with 55–60% of energy as carbohydrate is appropriate for all forms of hyperlipidemia and is a prudent diet for all adults (Bierman and Glomset, 1985). This dietary approach lowers serum lipids in hyperlipidemic individuals even more effectively than the AHA diet (Van Horn et al, 1988; Anderson et al, 1989b).

Patients like these high-fiber diets and, with adequate education, most report good to excellent adherence. Foods for the HCF diet are available locally and cost about 20% less than foods in a more traditional American diet (Anderson, 1985). Most individuals following an HCF diet note more flatulence but report no disturbing gastrointestinal symptoms. We have noted no adverse effects on vitamin or mineral status.

Effective management of hypercholesterolemia includes diet, exercise, weight reduction, and minimization of other risk factors (Anderson et al, 1990b). In patients with cholesterol levels above 350 mg/dl, pharmacological agents may be required (Anderson 1985). Because of side effects commonly associated with lipid-lowering drugs, however, nutrition and lifestyle therapy should be vigorously pursued before initiating pharmacologic agents.

Conclusions

In general, soluble fibers have important effects on glucose and lipid metabolism, whereas insoluble fibers have greater effects on intestinal transit time and stool production. Soluble fibers act rapidly to lower serum cholesterol levels, with maximum responses occurring after about 11 days in metabolic ward studies and after about four weeks in ambulatory studies (Anderson et al, 1990a).

Oat bran products, which are rich sources of water-soluble fiber, have more potent cholesterol-lowering action than virtually any other widely used food product. Diets supplemented with oat bran significantly reduce serum cholesterol concentrations of hypercholesterolemic individuals in both metabolic ward (Table I) and ambulatory studies (Table III). Oat bran feeding significantly and selectively reduces atherogenic LDL cholesterol levels in serum but maintains or increases protective HDL cholesterol levels.

Proposed mechanisms by which soluble fibers lower serum cholesterol are still under intense study; multiple mechanisms may interact to produce this effect. Oats and other sources of soluble fiber may lower serum cholesterol by altering bile acid secretion and metabolism, by modifying lipid and lipoprotein metabolism, by altering concentrations of SCFAs and hormones, or by a combination of these effects.

While exact mechanisms for the hypocholesterolemic effect of soluble fiber remain unclear, this effect has been well documented and is clinically significant. For every 1% reduction in serum cholesterol, there is a 2% reduction in the risk of CHD (Lipid Research Clinics Program, the Lipid Research Clinics Coronary Primary Prevention Trial, 1984). Thus, a 10–15% reduction in serum cholesterol related to oat bran would reduce estimated risk for coronary heart disease by 20–30%.

Hyperlipidemia, hypertension, diabetes, and obesity are major risk factors for CHD, and all are diet related. Since dietary fiber lowers blood pressure, increases insulin sensitivity, improves diabetes control, and aids in weight management (Anderson and Gustafson, 1988), increasing dietary fiber intake is emerging as one of the most important nutritional interventions for reducing risk of cardiovascular disease. Wide-spread adoption of diets high in carbohydrates and fiber, and use of palatable and inexpensive high-fiber foods such as oat bran, may significantly lessen hypercholesterolemia and cardiovascular disease in Western countries.

LITERATURE CITED

ANDERSON, J. W. 1985. Hyperlipidemia and diabetes: Nutrition considerations. Pages 133-159 in: Nutrition and Diabetes. L. Jovanovic and C. Peterson, eds. Alan R. Liss, New York.
ANDERSON, J. W., and BRIDGES, S. R. 1988. Dietary fiber content of selected foods. Am. J. Clin. Nutr. 47:440-447.
ANDERSON, J. W., and CHEN, W.-J. L. 1986. Cholesterol-lowering properties of oat products. Pages 309-333 in: Oats: Chemistry and Technology. F. H. Webster, ed. Am. Assoc. Cereal Chem., St. Paul, MN.
ANDERSON, J. W., and GUSTAFSON, N. 1988. Hypocholesterolemic effects of oat and bean products. Am. J. Clin. Nutr. 48:749-753.
ANDERSON, J. W., and SIESEL, A. E. 1990. Hypocholesterolemic effects of oat products. Pages 17-36 in: New Developments in Dietary Fiber, Physiological, Physiochemical and Analytical Aspects. I. Furda and C. J. Brine, eds. Plenum Press, New York.
ANDERSON, J. W., and WOOD, C. L. 1990. Oat bran and serum cholesterol. (Letter to the editor) New Engl. J. Med. 322:1747-1748.
ANDERSON, J. W., STORY, L., SIELING, B., and CHEN, W.-J. L. 1984a. Hypocholesterolemic effects of high-fiber diets rich in water-soluble fibers. J. Can. Diet. Assoc. 45:140-149.
ANDERSON, J. W., STORY, L., SIELING, B., CHEN, W.-J. L., PETRO, M.

S., and STORY, J. 1984b. Hypocholesterolemic effects of oat bran or bean intake for hypercholesterolemic men. Am. J. Clin. Nutr. 40:1146-1155.

ANDERSON, J. W., BRIDGES, S. R., SMITH, S. F., DILLON, D. W., and DEAKINS, D. 1989a. Oat bran increases serum acetate of hypercholesterolemic men. Clin. Res. 37:931A.

ANDERSON, J. W., GARRITY, T. F., SMITH, B. M., and WHITIS, S. E. 1989b. A comparison of the effects of high fiber-high carbohydrate and American Heart Association diets on lipid levels. Circulation 80(Suppl. 4):85.

ANDERSON, J. W., DEAKINS, D. A., and BRIDGES, S. R. 1990a. Hypocholesterolemic effects of soluble fibers; Mechanisms. Pages 339-363 in: Dietary Fiber: Chemistry, Physiology, and Health Effects. D. Kritchevsky, C. Bonfield, and J. W. Anderson, eds. Plenum Press, New York.

ANDERSON, J. W., DEAKINS, D. A., FLOORE, T. L., SMITH, B. M., and WHITIS, S. E. 1990b. Dietary fiber and coronary heart disease. Crit. Rev. Food Sci. Nutr. 29:95-147.

ANDERSON, J. W., SPENCER, D. B., HAMILTON, C. C., SMITH, S. F., TIETYEN, J., BRYANT, C. A., and OELTGEN, P. 1990c. Oat bran cereal lowers serum total and LDL cholesterol in hypercholesterolemic men. Am. J. Clin. Nutr. 52:495-499.

ANDERSON, J. W., GILINSKY, N. H., DEAKINS, D. A., SMITH, S. F., O'NEAL, D. S., DILLON, D. W., and OELTGEN, P. R. 1991a. Lipid responses to oat bran versus wheat bran intake in hypercholesterolemic men. Am. J. Clin. Nutr. 54:678-683.

ANDERSON, J. W., ZEIGLER, J., DEAKINS, D., FLOORE, T. L., DILLON, D. W., WOOD, C. L., OELTGEN, P. R., and WHITLEY, R. J. 1991b. Metabolic effects of high-carbohydrate, high-fiber diets for insulin-dependent diabetic individuals. Am. J. Clin. Nutr. 54:936-943.

BEYNEN, A. C., BUECHLER, K. F., VAN DER MOLEN, A. J., and GREELEN, M. J. H. 1982. The effects of lactate and acetate on fatty acid and cholesterol biosynthesis by isolated rat hepatocytes. Int. J. Biochem. 14:165-169.

BHATHENA, S. J., AVIGAN, J., and SCHREINER, M. E. 1974. Effect of insulin on sterol and fatty acid synthesis and hydroxymethylglutaryl CoA reductase activity in mammalian cells grown in culture. Proc. Natl. Acad. Sci. USA 71:2174-2178.

BIERMAN, E. L., and GLOMSET, J. A. 1985. Disorders of lipid metabolism. Pages 1108-1136 in: Textbook of Endocrinology. J. D. Wilson and D. W. Foster, eds. W. B. Saunders Co., Philadelphia.

BLACKBURN, N. A., REDFERN, J. S., JARJIS, H., HOLGATE, A. M., HANNING, I., SCARPELLO, J. H. B., JOHNSON, I. T., and READ, N. W. 1984. The mechanism of action of guar gum in improving glucose tolerance in man. Clin. Sci. 66:329-336.

BOSSELLO, O., COMINACINI, L., ZOCCA, I., GARBIN, U., FERRARI, F., and DAVOLI, A. 1984. Effects of guar gum on plasma lipoproteins and apolipoproteins C-II and C-III in patients affected by familial combined hyperlipoproteinemia. Am. J. Clin. Nutr. 40:1165-1174.

CASTELLI, W. P., DOYLE, J. T., GORDON, T., HAMES, C. G., HJORTLAND, M. C., HULLEY, S. B., KAGAN, A., and ZUKEL, W. J. 1977. HDL cholesterol and other lipids in coronary heart disease: The cooperative lipoprotein phenotyping study. Circulation 55:767-772.

CHEN, W.-J. L., and ANDERSON, J. W. 1979. Effects of plant fiber in

decreasing plasma total cholesterol and increasing high-density lipoprotein cholesterol. Proc. Soc. Exp. Biol. Med. 162:310-313.

CHEN, W. L., and ANDERSON, J. W. 1986. Hypocholesterolemic effects of
soluble fiber. Pages 275-286 in: Dietary Fiber: Basic and Clinical Aspects.
G. V. Vahouny and D. Kritchevsky, eds. Plenum Press, New York.

CHEN, W.-J. L., ANDERSON, J. W., and GOULD, M. R. 1981. Effects of oat
bran, oat gum, and pectin on lipid metabolism of cholesterol-fed rats.
Nutr. Rep. Int. 24:1093-1098.

CHEN, W.-J. L., ANDERSON, J. W., and JENNINGS, D. 1984. Propionate
may mediate the hypocholesterolemic effects of certain soluble plant fibers
in cholesterol-fed rats. Proc. Soc. Exp. Biol. Med. 175:215-218.

CONNOR, W. E., and CONNOR, S. L. 1977. Dietary treatment of hyperlipidemia. Pages 281-326 in: Hyperlipidemia: Diagnosis and Therapy. B. M.
Rifkin and R. I Levy, eds. Grune and Stratton, New York.

CUMMINGS, J. H., POMARE, E. W., BRANCH, W. J., NAYLOR, C. P. E.,
and MacFARLANE, G. T. 1987. Short-chain fatty acids in human large
intestine, portal, hepatic, and venous blood. Gut 28:1221-1227.

DAVIDSON, M. H., DUGAN, L. D., BURNS, J. H., BOVA, J., STORY, K.,
and DRENNAN, K. B. 1991. The hypocholesterolemic effects of β-glucan in
oatmeal and oat bran. JAMA J. Am. Med. Assoc. 265:1833-1839.

DE GROOT, A. P., LUYKEN, R., and PIKAAR, N. A. 1963. Cholesterol-lowering effect of rolled oats. Lancet 2:303-304.

DEMARK-WAHNEFRIED, W., BOWERING, J., and COHEN, P. S. 1990.
Reduced serum cholesterol with dietary change using fat-modified and oat
bran supplemented diets. J. Am. Diet. Assoc. 90:223-229.

GOLD, K. V., and DAVIDSON, D. M. 1988. Oat bran as a cholesterol-reducing dietary adjunct in a young, healthy population. West. J. Med. 148:299-
302.

GOTTO, A. M. 1986. Interactions of the major risk factors for coronary heart
disease. Am. J. Med. 80(Suppl. 2A):48-55.

GOULD, M. R., ANDERSON, J. W., and O'MAHONY, S. 1980. Biofunctional
properties of oats. Pages 447-460 in: Cereals for Food and Beverages. G. E.
Inglett and L. Munck, eds. Academic Press, New York.

ILLMAN, R .J., TRIMBLE, R. P., SNOSWELL, A. M., and TOPPING, D. L.
1982. Daily variations in the concentrations of volatile fatty acids in the
splanchnic blood vessels of rats fed diets high in pectin and bran. Nut.
Rep. Int. 26:439-446.

ILLMAN, R. J., TOPPING, T. L., McINTOSH, G. H., TRIMBLE, R. P.,
STORER, G. B., TAYLOR, M. N., and CHENG, B.-Q. 1988.
Hypocholesterolemic effects of dietary propionate: Studies in whole ani-
mals and perfused rat liver. Ann. Nutr. Metab. 32:97-107.

ISAKSSON, G., LUNDQUIST, I., and IHSE, I. 1982. Effect of dietary fiber on
pancreatic enzyme activity in vitro. Gastroenterology 82:918-924.

JENKINS, D. J. A., LEEDS, A. R., GASSUL, M. A., WOLEVER, T. M. S.,
GOFF, D. V., ALBERTI, K. G. M. M., and HOCKADAY, T. D. R. 1976.
Unabsorbable carbohydrates and diabetes: Decreased post-prandial
hyperglycemia. Lancet 2:172-174.

JENKINS, D. J. A., WOLEVER, T. M. S., NINEHAM, R., SARSON, D. L.,
BLOOM, S. R., AHERN, J., ALBERTI, K. G. M. M., and HOCKADAY, T.
D. R. 1980. Improved glucose tolerance four hours after taking guar with
glucose. Diabetologia 19:21-24.

JUDD, P. A., and TRUSWELL, A. S. 1981. The effect of rolled oats on blood
lipids and fecal steroid excretion in man. Am. J. Clin. Nutr. 34:2061-2067.

KANNEL, W. B. 1983. High-density lipoproteins: Epidemiologic profile and risks of coronary artery disease. Am. J. Cardiol. 52:9B-12B.

KEENAN, J. M., WENZ, J. B., MYERS, S. R., RIPSIN C., and HUANG, Z. 1991. Randomized, controlled, crossover trial of oat bran cereal in hypercholesterolemic subjects. J. Fam. Pract. 33:600-608.

KESTIN, M., MOSS, R., CLIFTON, P. M., and NESTEL, P. J. 1990. Comparative effects of three cereal brans on plasma lipids, blood pressure and glucose metabolism in mildly hypercholesterolemic men. Am. J. Clin. Nutr. 52:661-666.

KHAW, K.-T., and BARRETT-CONNOR, E. 1987. Dietary fiber and reduced ischemic heart disease mortality rates in men and women: A 12-year prospective study. Am. J. Epidemiol. 126:1093-1102.

KIRBY, R. W., ANDERSON, J. W., SIELING, B., REES, E. D., CHEN, W.-J. L., MILLER, R. E., and KAY, R. M. 1981. Oat bran intake selectively lowers serum low density lipoprotein cholesterol concentrations of hypercholesterolemic men. Am. J. Clin. Nutr. 34:824-829.

LEADBETTER, J., BALL, M. J., and MANN, J. I. 1991. Effect of increasing quantities of oat bran in hypercholesterolemic people. Am. J. Clin. Nutr. 54:841-845.

LIPID RESEARCH CLINICS PROGRAM, THE LIPID RESEARCH CLINICS CORONARY PRIMARY PREVENTION TRIAL RESULTS. 1984. The relationship of reduction in incidence of coronary heart disease to cholesterol lowering. JAMA J. Am. Med. Assoc. 251:365.

MUNOZ, J. M., SANDSTEAD, H. H., and JACOB, R. A. 1979. Effects of dietary fiber on glucose tolerance of normal men. Diabetes 28:496-502.

POMARE, E. W., BRANCH, W. J., and CUMMINGS, J. H. 1985. Carbohydrate fermentation in the human colon and its relation to acetate concentrations in venous blood. J. Clin. Invest. 75:1448-1454.

REAVEN, G. M., and BERNSTEIN, R. M. 1978. Effect of obesity on the relationship between very low density lipoprotein production rate and plasma triglyceride concentration in normal and hypertriglyceridemic subjects. Metabolism 27:1047-1054.

REDARD, C. L., DAVIS, P. A., and SCHNEEMAN, B. O. 1990. Dietary fiber and gender: Effect on postprandial lipemia. Am. J. Clin. Nutr. 52:837-845.

SCHNEEMAN, B. O. 1982. Pancreatic and digestive function. Pages 73-83 in: Dietary Fiber in Health and Disease. G. V. Vahouny and D. Kritchevsky, eds. Plenum Press, New York.

SCHNEEMAN, B. O., and GALLAHER, D. 1985. Effects of dietary fiber on digestive enzyme activity and bile acids in the small intestine. Proc. Soc. Exp. Biol. Med. 180:409-414.

SCHNEEMAN, B. O., CIMMARUSTI, J., COHEN, W., DOWNES, L., and LEFEVRE, M. 1984. Composition of high density lipoproteins in rats fed various dietary fibers. J. Nutr. 114:1320-1326.

SCHWARTZ, S. E., LEVINE, R. A., SINGH, A., SCHEIDECKER, J. R., and TRACK, N. S. 1982. Sustained pectin ingestion delays gastric emptying. Gastroenterology 83:812-817.

STORCH, K., ANDERSON, J. W., and YOUNG, V. R. 1984. Oat-bran muffins lower serum cholesterol of healthy young people. Clin. Res. 34:740A.

STORY, J. A. 1985. Dietary fiber and lipid metabolism. Proc. Soc. Exp. Biol. Med. 180:447-452.

STORY, J. A., and KRITCHEVSKY, D. 1976. Comparison of the binding of various bile acids and bile salts in vitro by several types of fiber. J. Nutr. 106:1292-1294.

STORY, J. A., and LORD, S. L. 1987. Bile salts: In vitro studies with fiber components. Scand. J. Gastroenterol. 22(Suppl. 129):174-180.

STORY, J. A., and THOMAS, J. N. 1982. Modification of bile acid spectrum by dietary fiber. Pages 193-201 in: Dietary Fiber in Health and Disease. G. V. Vahouny and D. Kritchevsky, eds. Plenum Press, New York.

SWAIN, J. F., NOUSE, I. L., CURLEY, C. B., and SACKS, F. M. 1990. Comparison of the effects of oat bran and low-fiber wheat on serum lipoprotein levels and blood pressure. N. Engl. J. Med. 332:147-152.

TOPPING, D. L., ILLMAN, R. J., TAYLOR, M. N., and McINTOSH, G. H. 1985. Effects of wheat bran and porridge oats on hepatic portal venous volatile fatty acids in the pig. Ann. Nutr. Metab. 29:325-331.

TURNBULL, W. H., and LEEDS, A. R. 1987. Reduction of total and LDL-cholesterol in plasma by rolled oats. J. Clin. Nutr. Gastroenterol. 2:177-180.

VAHOUNY, G. V., TOMBES, R., CASSIDY, M. M., KRITCHEVSKY, D., and GALLO, L. 1980. Dietary fibers: V. Binding of bile salts, phospholipids and cholesterol from mixed micelles by bile acid sequestrants and dietary fibers. Lipids 15:1012-1018.

VAN HORN, L., LIU, K., PARKER, D., EMIDY, L., LIAO, Y., PAN, W. H., GIUMETTI, D., HEWITT, J., and STAMLER, J. 1986. Serum lipid response to oat product intake with a fat-modified diet. J. Am. Diet. Assoc. 86:759-764.

VAN HORN, L., EMIDY, L. A., LIU, K., LIAO, Y., BALLEW, C., KING, J., and STAMLER, J. 1988. Serum lipid response to a fat-modified, oatmeal-enhanced diet. Prev. Med. 17:377-386.

WRIGHT, R. S., ANDERSON, J. W., and BRIDGES, S. R. 1990. Propionate inhibits hepatic lipid synthesis. Proc. Soc. Exp. Biol. Med. 190:26-29.

INDEX

(T) indicates table only.

Acetate, and cholesterol
  synthesis, 123, 125, 149, 150
Adenocarcinoma, 129
Age, effect on serum cholesterol
  response, 147
Air classification, 30, 31, 39
Alcohol, and lipase activity, 34
Aleurone
  constituents, 11
  microchemical characteristics,
    10–13
American Heart Association (AHA)
  diet, 144, 145, 147, 152
Amino acid composition
  of oats and oat products, 57–60
  of other cereals, 61
  of soybean, 61
Amino acids, 10, 11, 57–61, 100
Aminophenol, 11
Amylase, 36, 62, 97
Apolipoprotein, 141, 143, 149
Apple fiber, 42
Arabinoxylan, 68, 71
Aromatic amines, 6(T), 10, 11
Ash, 1, 27, 53, 56, 62, 67
Atherosclerosis, 150–151
Autofluorescence, 17
Avenalumic acids, 10, 11
Avenanthramides, 10

Barley
  amino acids in, 61
  ash in, 53
  chick growth, effect on, 99–100
  dietary fiber in, 64(T), 70–71,
    72–73(T)
  ß-glucan in, 17, 65, 70, 106
    distribution, 20
    extractions, 92
    structure, 84–91

lipid in, 51(T)
oligosaccharides in, 55
protein in, 51
proximate composition, 51(T),
  52(T)
starch in, 54
sugars in, 54
water hydration capacity, 43(T)
Bile acids, 116, 123, 126, 140, 141,
  147–148
Bile salts in diet, 116
Blood glucose, 102–103
Bran
  milling of other cereals, 2
  morphology, 1–2, 9

Calcofluor, 6, 7, 8, 9, 15, 16, 95, 98
Carbohydrate content, 53–55
Carcinogenesis, 128–129
Cell proliferation, intestinal, 127,
  128, 129
Cell shape, 15
Cell size, 19
Cellulase, 90, 97
Cellulolytic enzymes, in processing,
  35–36
Cellulose, 68, 70, 71, 102, 124, 125,
  148, 149
Cell wall
  detection methods, 6(T)
  influence on cooking, digestion,
    and processing, 13
  thickness, 15, 16, 17, 18, 19, 20,
    92
Chenodeoxycholic acid, 148
Cholesterol
  in diet, 116
  inhibition of synthesis, 123-125,
    150

159

Gastric emptying, 100–101, 149
Gastrointestinal tract
  convective movements of, 127
  transit time, 101, 102
  unstirred layer, and diffusion,
    101, 127
  viscosity, 99, 100, 127
Gender, effect on serum cholesterol
    response, 147
Glucagon, 150
β-Glucan
  in aleurone, 11
  blood glucose, 102–103
  blood insulin, 102–103
  chick growth, 99–100
  collaborative study, 74
  concentration, 60(T), 64–65,
    68(T), 75(T), 76(T), 77(T), 78
  in barley and other grains,
    64–65, 70
  in gastrointestinal tract, 101
  in groat, 15, 65, 67
  in hull, 65
  in oat bran, 43, 65, 67, 71, 74,
    76
  in oat bran definition, 2
  conformation and shape, 87, 88,
    89
  in crease, 17
  detection methods, 6(T)
  distribution, 13–20
  in endosperm cell walls, 2
  enrichment in bran, 26, 27, 42
  extraction, 38, 39, 40, 91–93, 96,
    97
  gastrointestinal effects, 100–102
  glycemic response, 102–103
  interaction with Calcofluor, 6, 98
  linkage sequences, 84–90
  mapping, 8–9
  methods of analysis, 96–98
  molecular weight, 95
  peptide linkage, 91
  periodate oxidation of, 90
  physical properties, 91–95
  physiological effects, 98–106
  protein association, 90–91
  purification, 96

  serum cholesterol reduction, 103,
    105–106
  solubilase, 90
  solubility, 91–93
  structure, 7, 84–91
  viscosity, 93–95
  and yield of bran, 29
  X-ray crystallography, 87–88
β-Glucanase, 90, 96, 97, 99, 100
Glucose tolerance, 150
Glycemic effect, 102–103, 126–127
Granulation, of oat bran, 26, 29
Gravimetric analysis of dietary
  fiber, 61, 62, 63, 71–77
Groat structure, 9, 10
  distribution of β-glucan, 13,
    15–20
  methods for examination, 3–9
  microscopy and microchemistry,
    3–5
  microspectrophotometry, 6–8
Guar gum, 94, 95, 115, 116, 117

HCF diet, see High-carbohydrate
  high-fiber diet
HDL cholesterol, see High-density
  lipoprotein cholesterol
Heat treatment of seed, 28, 29
Hemicellulolytic enzymes, in
  processing, 35
High-carbohydrate high-fiber diet,
  140, 145, 151, 152
High-density lipoprotein
  cholesterol, 116, 139–140, 142,
    143, 146, 147, 149, 151, 152
Hulls
  amino acids, 58(T)
  dietary fiber in, 42, 53, 65, 67, 70,
    77, 78
  and processing, 28, 33, 42
  proximate analysis, 51, 52, 53,
    54, 55, 56, 57, 59, 60
  and serum cholesterol, 115, 117
Hyperlipidemia, 143, 151–152, 153
Hypocholesterolemic effect
  absorption effects, 125–126,
    148–149
  in animals, 114–126